AF622483

Integrated Deliberative Decision Processes for Water Resources Planning and Evaluation

Integrated Deliberative Decision Processes for Water Resources Planning and Evaluation

Edited by

Giorgos Kallis, Nuno Videira, Paula Antunes and Rui Santos

Publishing

Published by IWA Publishing, Alliance House, 12 Caxton Street, London SW1H 0QS, UK

Telephone: +44 (0) 20 7654 5500; Fax: +44 (0) 20 7654 5555; Email: publications@iwap.co.uk
Web: www.iwapublishing.com

First published 2007

British Library Cataloguing in Publication Data
A CIP catalogue record for this book is available from the British Library

Library of Congress Cataloging- in-Publication Data
A catalog record for this book is available from the Library of Congress

ISBN: 1843391678

ISBN 9781843391678

Contents

EDITORS AND PRINCIPAL AUTHORS

Giorgos Kallis, *Environmental Planning Laboratory, University of the Aegean, gkallis@env.aegean.gr*
Nuno Videira, *Ecoman, DCEA-FCT, New University of Lisbon, nmvc@fct.unl.pt*
Paula Antunes, *Ecoman, DCEA-FCT, New University of Lisbon, mpa@fct.unl.pt*
Rui Santos, *Ecoman, DCEA-FCT, New University of Lisbon, rfs@fct.unl.pt*

CONTRIBUTORS - ASSOCIATED AUTHORS (IN ALPHABETICAL ORDER)

Frank Bruinsma, *Free University of Amsterdam, Fbruinsma@feweb.vu.nl*
Harry Coccossis, *University of the Thessaly, hkok@prd.uth.gr*
Serafin Corral Quintana, *University of La Laguna, scorral@ull.es*
Ângela Guimarães Pereira , *EC – JRC – IPSC, Ispra , angela.pereira@jrc.it*
Leandro del Moral, *University of Seville, lmoral@us.es*
Dionisia Hatzilacou , *Environmental Consultant , kirisis@otenet.gr*
Gonçalo Lobo, *Ecoman, DCEA-FCT, New University of Lisbon, ggl@fct.unl.pt*
Pilar Paneque, *Pablo de Olavide University, ppansal@dhuma.upo.es*
Belen Pedregal Mateos, *University of Seville, bpedregal@us.es*
Clive L. Spash, *Geography & Environment Department, University of Aberdeen, c.spash@macaulay.ac.uk*
Kevin Urama, *The Macaulay Institute, University of Aberdeen, k.urama@macaulay.ac.uk*
Ron Vreeker, *Free University of Amsterdam, rvreeker@feweb.vu.nl*

TEXT EDITOR

Theoni Dede, *University of the Aegean, enpl@env.aegean.gr*

ADVISOR Project

Integrated Evaluation for Sustainable River Basin Governance

Advisor (Integrated Evaluation for Sustainable River Basin Governance) is a research project funded by the European Commission under the Energy, Environment and Sustainable Development RTD Programme.

Contract EVK1-CT-2000-00074

Project co-ordinator

Paula Antunes

Ecoman - Ecological Economics and Environmental Management Centre
Department of Environmental Sciences and Engineering
Faculty of Sciences and Technology
New University of Lisbon
Quinta da Torre
2829-516 Caparica
PORTUGAL

ADVISOR Partners

Ecoman Centre, DCEA-FCT/UNL, Portugal

University of the Aegean, Greece

EC - Joint Research Centre - IPSC, Ispra, Italy

Free University of Amsterdam, The Netherlands

Macaulay Institute, University of Aberdeen, United Kingdom

University of Seville, Spain

Instituto da Água, Portugal

Instituto del Água de Andaluzia, Spain

A.N.E.T.K.Y., Greece

Project Documentation:

The reports and further information on the ADVISOR Project may be accessed through the Website http://ecoman.dcea.fct.unl.pt/projects/advisor

Foreword

This Guidance Document is part of the ADVISOR ("Integrated Evaluation for Sustainable River Basin Governance") research project funded by the European Commission, under the "Energy, Environment and Sustainable Development" theme of the 5th Framework Research Programme.

The aim of ADVISOR is to improve the understanding of evaluation processes as part of river basin planning and management and to provide a framework supported by a toolkit for the conduct of integrated and participatory evaluations.

Work Package 1 of the project examined past water project or plan evaluation cases in five EU states and drew insights on the problems of past evaluation practices. Work Package 2 interpreted these past experiences from different analytical angles leading to a theory for an integrated evaluation process, emphasising issues of deliberation, multiple values, quality in the use of information and governance.

Work Package 3 moved from theory to practice. Different tools which could contribute to integrated evaluation processes were tested in experimental case applications. These included scenario workshops, mediated modelling supported by a quality assurance protocol, social multi-criteria evaluation, cost-effectiveness analyses and monetary valuation.

The present document is part of Work Package 4 of the project, which aimed at transferring the experience and lessons learned during the ADVISOR project to policy makers, contributing especially to the implementation process of the Water Framework Directive. An Integrated Deliberative Decision Process (IDDP) is proposed to be adopted as the platform to achieve integrated evaluations. This document explains and provides a step-by-step guidance on how to design and run such a process.

This document is directed to policy makers. Some theoretical reflection however is also provided, where this bears importance to practical implementation. References to theoretical literature have been avoided to a maximum extent to facilitate the "flow of the text". However, lists of further reading are provided in the end of each section for those who wish to learn more on the issues discussed or find more details about its implementation.

Acronyms

CBA *Cost Benefit Analysis*
CC *Consensus Conference*
CEA *Cost Effectiveness Analysis*
CIS *Common Implementation Strategy of the Water Framework Directive*
CJ *Citizens' Juries*
CVM *Contingent Valuation Method*
DMV *Deliberative Monetary Valuation*
EC *European Commission*
EIA *Environmental Impact Assessment*
EU *European Union*
IDDP *Integrated Deliberative Decision Process*
IMV *Integrated Monetary Valuation*
MM *Mediated Modelling*
MS *Member States*
NHP *National Hydrological Plan*
QAP *Quality Assurance Protocol*
RBD *River Basin District*
RBMP *River Basin Management Plan*
SMCE *Social Multi-criteria Evaluation*
VW *Visioning Workshops*
WFD *Water Framework Directive*
WTA *Willingness to Accept*
WTP *Willingness to Pay*
CBA *Cost Benefit Analysis*
CC *Consensus Conference*
CEA *Cost Effectiveness Analysis*
CIS *Common Implementation Strategy of the Water Framework Directive*
CJ *Citizens' Juries*
CVM *Contingent Valuation Method*

Glossary

Citizens' Juries/ Consensus Conferences	*Tools for community opinion while addressing controversial issues. They are based on the rationale that given adequate information and opportunity, all people, even if not professionally trained, can be trusted to take decisions regarded as legitimate and fair on behalf of the community.*
Common Implementation Strategy	*Multi-stakeholder approach of the EU to support the implementation of the Water Framework Directive.*
Deliberation	*A careful consideration or discussions of reasons for and against (Oxford English Dictionary).*
Deliberative decision	*A decision that involves social interaction, language-based discussion and debate, respect, reflection, and transformation of different values and viewpoints (Holmes and Scoones, 2000).*
Evaluation	*Refers to the process of assessing and comparing the relative values (in the broadest sense) of one or more courses of action.*
Governance	*Governance refers to the process whereby elements in society wield power and authority, and influence and enact policies and decisions concerning public life and sustainability. It is a broader notion than government, whose principal elements include the constitution, legislature, executive and judiciary. Governance involves interaction between these formal institutions and those of civil society (IIAS, 1999).*
Integrated Deliberative Decision Process	*A deliberative decision process consisting of one or more debates with different participants' representations, based on social robust information and knowledge; taking explicitly into account in its design and conduct the presence of multiple values and being embedded in an inclusive and discursive governance framework (institutional, regulatory, social).*
Mediated Modelling	*A tool for overcoming problems inherent in linear thinking and compartmentalized, non-participatory decision making. It aims for a collaborative team learning experience to raise the shared level of understanding in a group, as well as fostering a broad and deep consensus (van den Belt, 2004).*
Monetary valuation	*Estimation in monetary terms of the value of environmental services and goods.*
Multi-criteria evaluation	*A decision aid method based on the analysis of a number of alternatives, taking into account conflicting interests and multiple criteria, usually including economic, social and environmental factors.*
Public participation	*In broad terms refers to any of the several mechanisms (processes and tools) intentionally designed for involving lay public and/or stakeholders (the two groups together are often referred to as interested parties) in decision-making.*
Quality Assurance	*In environmental policy related issues, the maintenance of quality depends on an open dialogue among all those affected, where quality is understood as a contextual property of scientific information. The involvement of the relevant social actors (stakeholders, etc.) is suggested as a transversal element to ensure the quality of the whole evaluation procedures in river basin governance.*
River Basin	*The area of land from which all surface run-off flows through a sequence of streams, rivers and, possibly, lakes into the sea at a single river mouth, estuary or delta.*
Visioning Workshop	*Workshop which allows for creating a desired vision about the future of the examined system and generating, ranking and designing the implementation of ideas for its realisation.*

EXECUTIVE SUMMARY

This guidance document provides a process and tools for directly involving the various stakeholders and the public in decision aid and decision-making in water resource planning and management. The platform for this involvement is an **Integrated Deliberative Decision Process (IDDP)**.

A deliberative decision process is one in which decision is the outcome of **interaction through dialogue** between a broad and inclusive range of people with respect for their different viewpoints. An IDDP: combines several tools and participation events in a longer-term process; caters for a democratic use of scientific information; recognises the presence of multiple values; is aware of its inter-relationship with the broader socio-political ("governance") framework.

The need for direct public involvement in water decision-making (planning, project evaluations, etc) has been recognised in several international forums. The new EU Water Framework Directive gives a new impetus to participation with its mandate for the active involvement of the public in river basin planning formulation and implementation. Planning, evaluation and participation however have been dealt with as separate tasks in the Directive. The present document addresses this gap providing a practical platform for directly involving the public and stakeholders in the formulation of plans or the evaluation of projects.

The target audience of this document includes those authorities in the Member States of the European Union responsible for decisions in water resource management. This includes: competent authorities of river basin districts, other national, regional or local agencies with responsibilities in water resource management or the authorization of water resource projects and water utilities. Different elements of the document will also be interesting for European Commission DGs Environment, Regions and Research.

An **IDDP** includes seven basic steps:

(1) Setting up of the organisational scheme
An institutional actor should undertake the responsibility for the whole process and build up a partnership with main stakeholder and citizen groups. A Steering Group of representatives, assisted by professional groups where relevant should then cater for the planning and implementation of the IDDP.

(2) Definition of decision and deliberation goals
The issues and questions addressed (i.e. the decision) and the expectations from deliberation should be debated by the partnership and decided at an early stage of the process. Deliberation goals might stem from exchanging information or educating the participants to resolving conflicts of agreeing in common plans and projects.

(3) Designing deliberative tools and participation process
Tools should then be selected to fulfil the defined deliberation and decision goals. One or more tools may be used and combined, depending on purposes. The Advisor project has tested three basic techniques: visioning workshops, mediated modelling workshops and social multi-criteria evaluation. Together with deliberative monetary valuation and citizens' juries/consensus conferences, these provide a basic (though not exhaustive) toolkit that can serve basic water planning and decision needs. Broader participatory processes such as public hearings, opinion polls, referendums, etc should complement stakeholder or random group deliberation processes.

(4) Establishment of an information Quality Assurance Protocol
Clear guidelines and rules, formalised in a protocol, should be used to guide the management of scientific information for the deliberative decision processes.

(5) Planning of event(s)
The process and the individual deliberative decision events (workshops, group meetings, etc) should be planned carefully well in ahead. Sequential tasks include: preparation of a scoping document with title and expected outcomes of the process; preparation of a detailed work plan with an accompanying time table presenting activities to be carried out, events to be held and participants to be involved; formulation of a detailed budget for the process indicating allocation per category of costs and sources of funding; preparation of specific work plans for the deliberative events to be carried out.

(6) Implementation of event(s)
This will depend on the tools(s) used. Professional facilitation is crucial for deliberative decision meetings. In this document practical guidance on how to implement the three tools of the Advisor toolkit is provided.

(7) Evaluation of the process and its results
Results of the process should be evaluated both in terms of the quality of the procedure and in terms of contribution to the decision goals. Evaluation should be performed by the organisers together with the partnership of the stakeholders involved and the broader community.

Deliberative decision processes have limitations. Framing of deliberation goals should be done with caution and should be inclusive, as it is decisive for the nature and results of the process. Design of the process can decide outcomes; the process itself should be adequately deliberated in advance. More than one tools and types of representation/participation should be used together to overcome the limited legitimacy of any given representation form alone. Note that deliberative decision-making is not proposed here as an alternative but as a complement to legitimate representative democracy structures.

An IDDP will be successful only if it really matters. Mechanisms should be established for embedding the outcomes of deliberative processes in real decision-making.

An IDDP is conditioned by the constraints set by the surrounding social, institutional and power relationships. Still, reciprocally deliberative processes can build up an impetus for changes in the broader social context.

Power inequalities and exclusions in the social sphere will invariably lead to inequalities in the deliberative process unless explicitly mitigated. Participants with more resources, information or ability to express and articulate ideas will be privileged in the deliberative process. An inclusive and transparent governance framework is necessary for a successful IDDP. Clear regulation of the rules and entitlements of a deliberative process should be set.

Introduction

A Guidance Document: What For?

THE PURPOSE OF THIS GUIDANCE DOCUMENT

This document focuses on how to design and run an **integrated deliberative process for supporting decision-making** in water resource planning and evaluation, especially in the broader context of implementing the **Water Framework Directive (WFD)** (Directive 2000/60/EC, of 23 October 2000).

Integrated Deliberative Decision Processes (IDDPs) are useful for different steps and phases in water resource planning and evaluation and apply to most water management activities from drafting a **river basin plan**, to the designation of a **protected area** or the decision to proceed or not with the construction of **a dam**.

To whom is this document addressed?

This document is primarily addressed to those **authorities** in the Member States (MS) of the European Union (EU) responsible for decisions in water resource management, including:

- Competent authorities of **river basin districts**.
- Other national, regional or local **agencies** with responsibilities in water resource management or the authorization of water resource projects.
- **Water utilities**.

The EU MS, Norway and the European Commission (EC) have jointly developed a **Common Implementation Strategy (CIS)** for supporting the WFD involving the preparation of implementation guidance by specialized working groups. The present document complements the work of these groups.

The document is also directly useful to stakeholders and the broader public. It informs them on the need for IDDPs and their requirements and it equips them with a procedural blueprint for a sound participatory decision process against which on-going ones (e.g. river basin plans, authorizations of contentious projects, etc) can be compared.

EU institutions involved in the implementation of the WFD (**DG-Environment**) or in evaluations of water resource related projects (**DG-Regio**) will find in the present document a useful blueprint for sound evaluations with direct involvement of the public. The last section of the document will also be of interest to **DG-Research**, as it identifies future research needs.

What can you find in this document?

This document:

- Aims at creating a **common understanding** of IDDPs and its benefits for water resource planning and management, especially in the context of the WFD;
- Provides guidance on **how to** design and implement a comprehensive IDDP for water resource planning or specific decisions;
- Provides **tools** and examples of application in order to make IDDPs operational.

What you will not find in this document

This is not a manual on IDDP. Whereas general principles, procedures and available tools for designing and running an IDDP will be provided, actual implementation of the process and design of the tools should be tailored to the needs of the specific decision/application and to its political, organizational, cultural and environmental context.

This document does not focus on providing detailed guidance on how to apply specific deliberative decision tools although some basic information is provided in the toolkit section. Detailed information about the application of specific tools should be sought to the specialized literature to which the reader is referred to appropriately in this text.

Where to find additional and updated information

The full ADVISOR project reports are available for download from the website http://ecoman.dcea.fct.unl.pt/projects/advisor.

A specific website on the IDDP guidance, including further information on the toolkit and its application to the ADVISOR case studies is also available at http://ecoman.dcea.fct.unl.pt/projects/advisor/iddp.

What does this document add to existing guidance?

Three documents relevant to the present guidance have been produced as part of the WFD-CIS.

Guidance documents of the WFD CIS related to present guidelines
(http://europa.eu.int/comm/environment/water/water-framework/guidance_documents.html)

Guidance on Public Participation in relation to the Water Framework Directive. Active involvement, Consultation and Public Access to information. Produced by Working Group 2.9 - Public Participation.

Economics and the Environment. The Implementation Challenge of the Water Framework Directive. A Guidance Document. Produced by Working Group 2.6 – WATECO.

Guidance on Planning Processes. Produced by Working Group 2.9 – Planning Processes

The present document:

a) Connects existing WFD guidance.
The aforementioned guidance documents address plan formulation, evaluation (economic) and participation as three separate tasks. Deliberative processes build a "bridge" between participation and decision-making providing tools for the active involvement of the public in the formulation of plans or the evaluation of projects.

b) Exceeds the scope of the implementation process of the WFD.
In addition to the formulation / implementation of river basin plans and decision-making in the context of the WFD, the present guidance is also relevant to "quantitative" water plans and decisions (e.g. water supply and demand plans, allocation of water resources, authorization of new waterworks) and urban water utility planning and management.

Chapter 1

Water Resource Planning, Evaluation and the Water Framework Directive: Setting the Scene

WATER RESOURCE MANAGEMENT

Water resource management refers to activities that aim to coordinate humans´ goals with the conditions (ecological, hydrological) of water systems. Water management includes **physical measures** such as:

- Impoundments (dams, reservoirs, etc) and diversions (transfers) typically for water supply or energy production;
- Wells for groundwater abstraction (or for groundwater injection and aquifer recharge);
- River regulation works (channelisation, etc) typically for navigation or flood protection;
- River restoration projects for ecological, aesthetic or recreational purposes;
- Pollution control technologies.

and softer, **non-physical instruments** such as:

- Regulatory controls (e.g. emission limit standards, designation of protected areas and control of certain activities within boundary zones, etc);
- Economic measures (e.g. pricing of water use, taxing of water abstractions, etc);
- Measures aimed at the civil society (e.g. programmes for public awareness for water saving).

WATER RESOURCE PLANNING

Water resource planning is a rational, staged process of selecting and implementing the best mix of water management measures and instruments in order to achieve defined goals. It involves a process starting with a scoping of the problems to be addressed and collection of data to appraise conditions. This is followed by the definition of the goals that the plan should fulfil the elaboration of the goals into a set of criteria upon which to assess alternative options and the identification of alternative management measures and instruments. Alternatives are then comparatively evaluated upon the identified criteria leading to the compilation of an optimal mix of measures and instruments in a programme or action plan. Continuous collection of data and monitoring of the results of the measures and instruments accompany the plan's implementation. New problems and needs arise leading to a new cycle of planning. This should be seen as a continuous, iterative process, the different stages being inter-related, rather than a "one-through" clearly defined process of consecutive steps.

Water resource planning requires that quantity and quality issues are addressed together, that a full range of options is taken into account (physical and non-physical) and that the evaluation includes multiple criteria (technical feasibility, cost, environmental, social, etc).

Figure 1.1 portrays such generic planning process applicable at different spatial and organisational levels. A national agency can prepare a water resource plan at the national level and a municipal authority or water utility at the local, urban or metropolitan level. The river basin[1] provides an appropriate territorial unit for water resource planning, accounting for the interdependency of water quantity and quality, water and adjacent land-based resources, and upstream and downstream effects.

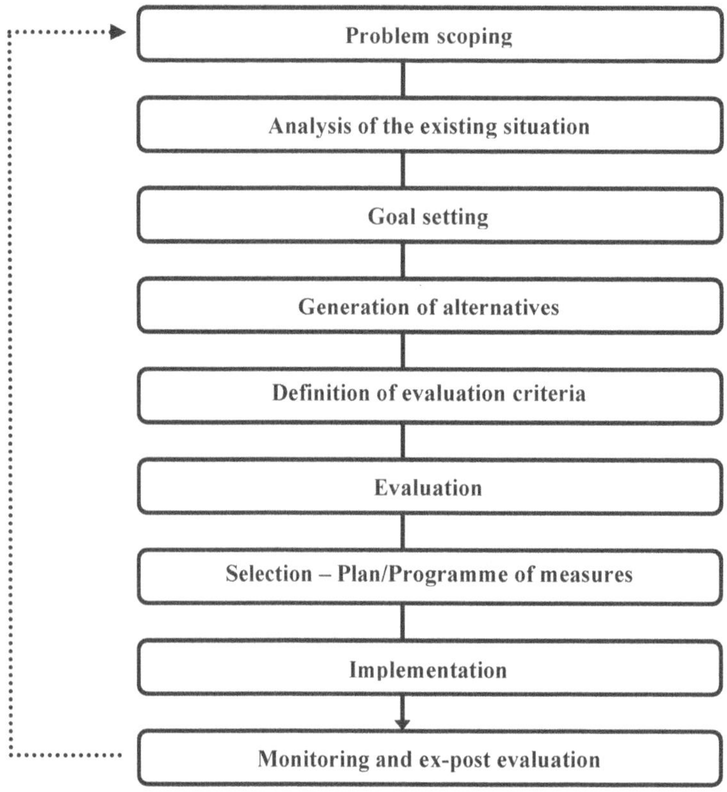

Figure 1.1 Generic water resource planning process

Further reading on Water Resource Planning and Management

Global Water Partnership (2000), Integrated Water Resources Management, *TAC Background Papers no 4*, Technical Advisory Committee, Global Water Partnership, Stockholm.

Grigg, N.S. (1996), *Water Resources Management – Principles, Regulations and Cases*, McGraw-Hill, New York.

EVALUATIONS IN WATER RESOURCE PLANNING AND MANAGEMENT

Evaluation refers to the process of assessing and comparing the relative values (in the broadest sense) of one or more courses of action. Evaluation and comparison of the advantages and disadvantages of the different management options is a crucial stage of the planning process, as depicted in Figure 1.1.

In reality many water management decisions are taken outside a formal planning process. Authorities have to decide whether to authorize or not a new water abstraction or dam, proceed or not with the restoration of a polluted river, designate or not an area as protected, or how to design new water tariffs. Ideally, all these decisions should be addressed as part of a comprehensive planning process. In practice, there will always be a need for evaluations of individual projects, policies, etc.

PARTICIPATION IN WATER RESOURCE PLANNING AND MANAGEMENT

Public participation is now broadly recognised as a key requirement of water resource planning and management. The **Dublin Statement on Water and Sustainable Development**, a milestone in international water policy[2] recognised as

[1] A river basin is defined in the WFD as "the area of land from which all surface run-off flows through a sequence of streams, rivers and, possibly, lakes into the sea at a single river mouth, estuary or delta" (Directive 2000/60/EC, Official Journal L 327/1, 22.12.2000).

one of its four key principles the need for "a participatory approach involving users, planners and policy-makers at all levels". It called for raising awareness of the importance of water among policy-makers and the general public and for taking decisions at the lowest appropriate level, with full public consultation and involvement of users in the planning and implementation of water projects.

PLANNING IN THE WATER FRAMEWORK DIRECTIVE

Box 1.1 summarises the main provisions of the WFD. The Directive introduces a new planning process at the river basin level (Figure 1.2). The outcome is a river basin plan. The goal is the achievement of the environmental objectives of the Directive for the waters of the river basin district. The process follows the basic logic of the planning process described in Figure 1.1. Issues are identified, objectives are defined, alternative measures are evaluated and a cost-effective programme is designed and implemented. Monitoring and ex-post evaluation follow, identifying continuing gaps and leading to a reformulation of the plan and the programme of measures.

The WFD river basin planning process is only a subset of a broader water resource planning process. Quantitative aspects are addressed in the WFD only to the extent that they affect the achievement of the environmental objectives of the Directive. Some MS may opt and combine resource/quantitative planning processes (e.g. decisions for water allocation, new waterworks, demand management programmes, etc) with the WFD river basin planning process. Others might not. Furthermore, the Directive allows for additional plans at the sub-basin level, specific issues or sectors. Water utilities or irrigation organisations for example, will continue having their own planning processes for their systems, in parallel to the WFD planning process.

EVALUATIONS IN THE WATER FRAMEWORK DIRECTIVE

Evaluations will be needed both in the river basin planning process and for specific decisions thereof, such as for the designation of protected areas or heavily modified waters and for decisions on water bodies that will be exempted from the requirement for the good status (Box 1.2).

In addition, the WFD requires that all uses or other activities that interfere with the aquatic systems and are suspected of having "significant adverse impacts on the status of water" (e.g. a new dam or a new road infrastructure) are subject to prior authorization upon compliance with the objectives of the Directive. An exemption is foreseen for "new modifications" to the physical characteristics of surface water or groundwater and "new sustainable human activities" (generally taken to refer to new dams for drinking water supply or electricity production, river regulation for navigation, etc) <u>if</u>:

a) the reasons are of "overriding public interest" and/or the benefits to the environment and to society of achieving the objectives are "outweighed" by the benefits of the new modifications or alterations to human health, to the maintenance of human safety or to sustainable development <u>and</u>,
b) the beneficial objectives served by those modifications or alterations of the water body cannot for reasons of technical feasibility or disproportionate costs be achieved by other means which are a significantly better environmental option. In these cases it suffices to take only practical mitigation measures and to explain and justify the reasons in the river basin plans.

Evaluation thus holds a key position in the WFD implementation process. Yet there is not direct guidance or support to river basin authorities on how to carry out these critical evaluations.

---

[2] The Dublin conference was attended by five hundred participants, including government-designated experts from a hundred countries and representatives of eighty international, intergovernmental and non-governmental organizations. The Dublin Statement was addressed the world leaders assembled at the United Nations Conference on Environment and Development (UNCED) in Rio de Janeiro in June 1992, and has provided the basis for following international water policies and agreements.

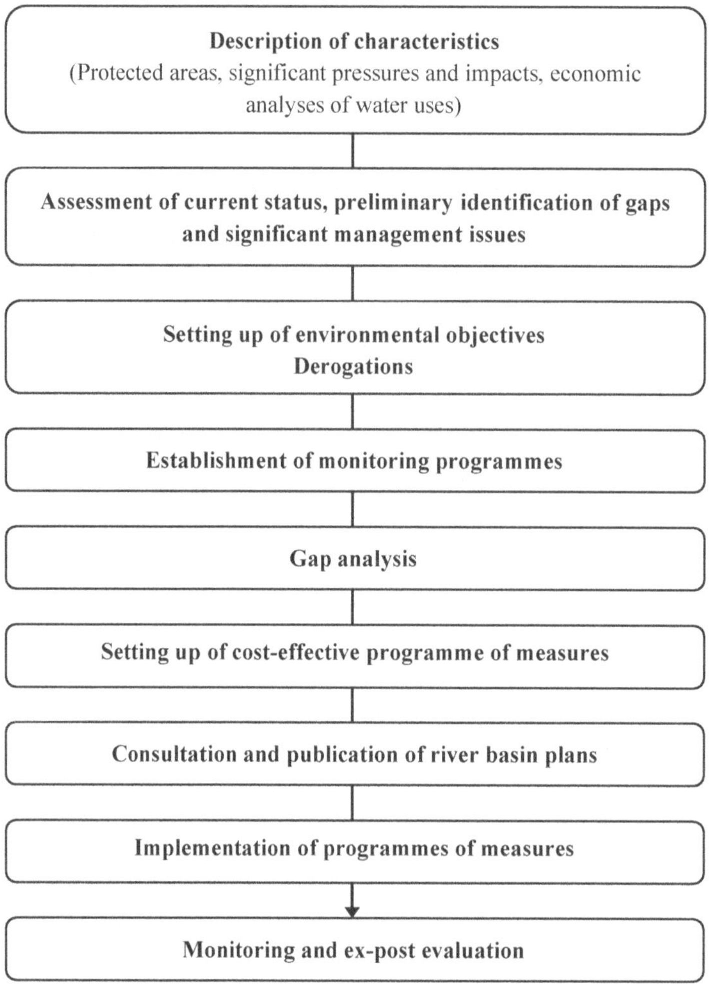

Figure 1.2 The WFD River Basin Planning process *(After Guidance on Planning Processes,* C.I.S Working Group 2.9)

The only relevant guidance on evaluation is this from the **Working Group on "Economics and the Environment"**. Certainly, economic valuation techniques are important in evaluation. As the Working Group itself however recognises in its guidance document, economic analysis is only "a process of providing valuable information to aid decision making". While economic valuation tools are an essential part of the overall approach for supporting decisions, they do not stand alone. There are many important values, other than monetary, which cannot be grasped by economic techniques alone.

Indeed, the wording of the Directive with terms such as "other means", "significant effects", "sustainable activities", "feasibility", "disproportionate costs", "reasonably be achieved", "significantly better environmental option", "environmental and socioeconomic needs", expands the range of evaluation needed and the dimensions to be taken into account. The Working Group in Economics recognizes that a judgment on "disproportional costs"... encompasses many complex decisional, institutional and socio-economic elements ... [and] ... will require co-ordination with other experts and **consultation of key stakeholders and the public**".

Box 1.1 The EU Water Framework Directive (2000/60/EC, Official Journal L 327/1, 22.12.2000)

The WFD sets a common legal framework for the protection of all waters (including inland surface waters, transitional waters, coastal waters and groundwater) which:

- Prevents further deterioration of, protect and enhance the status of water resources;
- Promotes sustainable water use based on long-term protection of water resources;
- Aims at enhancing protection and improvement of the aquatic environment through specific measures for the progressive reduction of discharges, emissions and losses of priority substances and the cessation of phasing-out of discharges, emissions and losses of the priority hazardous substances;
- Ensures the progressive reduction of pollution of groundwater and prevents its further pollution; and
- Contributes to mitigating the effects of floods and droughts.

The core environmental objective of the Directive is the achievement of a **"good" status** for all waters in terms of ecological, chemical and quantitative (for groundwater only) parameters within 15 years (with possibility for 12 years extension) and to prevent any further deterioration of aquatic ecosystems.

The **river basin** is the spatial unit upon which to plan and coordinate measures for the achievement of objectives. MS have to designate authorities responsible for the preparation and implementation of river basin plans and **programmes of measures**. The programme should include existing legally-set instruments (e.g. secondary sewerage treatment, good agricultural practices in nitrate sensitive zones, industrial installation permits) plus a number of mandatory new instruments defined in the WFD (costing/pricing, zoning of designated areas, abstraction and discharge permitting and authorisation of activities that impact water quality). If the aforementioned "basic measures" do not suffice to achieve the environmental objectives of the Directive, additional measures should be implemented by river basin authorities, such as applying stricter permit standards, zones for good farming practices beyond the nitrate sensitive areas, implementing water demand management programmes, etc.

The WFD also strengthens a **combined approach** to the control of pollutants, provides the public with rights to information and **participation** in decisions, and establishes a detailed system of monitoring and reporting.

The key actions that Member States need to take include:

- To identify the individual river basins lying within their national territory and assign them to individual River Basin Districts (RBDs) and identify competent authorities by 2003;
- To characterize river basin districts in terms of pressures, impacts and economics of water uses, including a register of protected areas lying within the river basin district, by 2004;
- To carry out, jointly and together with the EC, the intercalibration of the ecological status classification systems by 2006;
- To make operational the monitoring networks by 2006;
- Based on sound monitoring and the analysis of the characteristics of the river basin, to identify by 2009 a cost-effective programme of measures for achieving the environmental objectives of the Directive;
- To produce and publish River Basin Management Plans (RBMPs) for each RBD including the designation of heavily modified water bodies, by 2009;
- To implement water pricing policies that enhance the sustainability of water resources by 2010;
- To make the measures of the programme operational by 2012;
- To implement the programmes of measures and achieve the environmental objectives by 2015.

PARTICIPATION IN THE WATER FRAMEWORK DIRECTIVE

Participation in the river basin planning process is a key requirement of the WFD. The Directive requires explicitly that (at least) the significant water management issues and the draft river basin plans are available for comments to the public (for a period of 6 months to comment in writing). A timetable and work programme for the production of the plan, including a statement of the consultation measures to be taken, should also be made available at least three years before the beginning of the period to which the plan refers. Then the final river basin management plan shall include a summary of the public information and consultation measures taken, their results and the changes to the plan made as a consequence.

Beyond consultation, the Directive calls for the **active involvement** of all interested parties in its implementation, in particular in the production, review and updating of the RBMPs. A Working Group on Public Participation has been set up producing guidance on how to run consultation and participatory processes during the river basin planning process.

Box 1.2 Evaluation in the implementation of the WFD

Programme of measures
Different measures that can contribute to the achievement of the objectives should be evaluated and a "cost-effective" mix of measures be designed. Evaluation might also be needed to compare different possible "designs" of one measure (e.g. a river restoration project, a designation of a protected area).
Designation of protected areas
A register of protected areas is required. These include areas around drinking water sources, areas with "economically significant aquatic species", bodies of recreational waters including bathing water, nutrient-sensitive vulnerable zones and areas for the protection of habitats or species (including relevant Natura 2000 sites), where water is an important factor in their protection. Designation of such areas, their borders and applying restrictions (where this has not been done as part of existing Community legislation) will need an evaluation of the advantages and disadvantages of different spatial borders, measures implemented and limitations imposed on the different activities.
Designation of heavily modified waters
Modified waters generally refer to regulated rivers, dams or artificial canals. Less stringent objectives are applicable for these waters. The criterion by which a water body can be designated as heavily modified is an evaluation of whether the necessary hydromorphological changes for achieving good ecological status would have "significant adverse effects" on the wider environment, navigation, water storage and regulation, flood protection and other equally important "sustainable human development activities" and whether the "beneficial objectives" served by the artificial or modified characteristics of the water body cannot, for reasons of "technical feasibility" or "disproportionate costs reasonably be achieved by other means", which are a "significantly better environmental option".
Derogations from the objectives of the Directive
MS may aim to achieve less stringent environmental objectives for specific bodies of water when they are so affected by human activity that the achievement of the objectives would be "infeasible" or "disproportionately expensive" and the "environmental and socioeconomic needs" served by such human activity can not be achieved by other means, which are a "significantly better environmental option not entailing disproportionate costs". This requires an evaluation of action vs. no action costs and a multi-dimensional evaluation of the polluting human activity vs. alternatives in terms of the needs served plus costs and environmental impacts.

Further reading on the Water Framework Directive

Barreira, A. and G. Kallis (2003), The EU Water Framework Directive and Public Participation in transboundary river basin management, In: Timmerman J.G. and S.Langaas (Eds.), The role and use of environmental information in European transboundary river basin management, 92-107, London: International Water Association Publishing.

European Commission (2001), Strategic Document. Common Strategy on the Implementation of the Water Framework Directive.

Kallis, G. and Butler, D. (2001), "The New EU Water Framework Directive: Measures and Implications", *Water Policy*, vol. 3(3), pp.125-142.

Lanz, K. and S.Scheuer (2001), EEB Handbook on EU Water Policy under the Water Framework Directive, Brussels: European Environment Bureau.

C.I.S Working Group documents on Planning, Participation and Economics available at http://europa.eu.int/comm/environment/water/water-framework/guidance_documents.html

Rationale for the present guidance document

Whereas guidance has been provided for participatory processes, there has been no sufficient guidance on how to link them with actual decision-making and especially evaluations in the planning process. This linkage can be achieved by IDDPs and accompanying tools.

Chapter 2

Introducing Integrated Deliberative Decision Processes for Water Resources Planning and Evaluation

WHAT IS AN IDDP?

An integrated deliberative decision-making process is one:

- In which decision is the outcome of an explicit process of **deliberation** (assisted by relevant tools) with broad and inclusive participation of stakeholders and the public;
- Based on **socially robust information and knowledge;**
- That takes explicitly into account in its design and conduct the presence of **multiple values**;
- Embedded in an inclusive and discursive **governance** framework (institutional, regulatory, and social).

The tetrahedron of Figure 2.1 illustrates the interrelation between the various factors that are necessary for an IDDP.

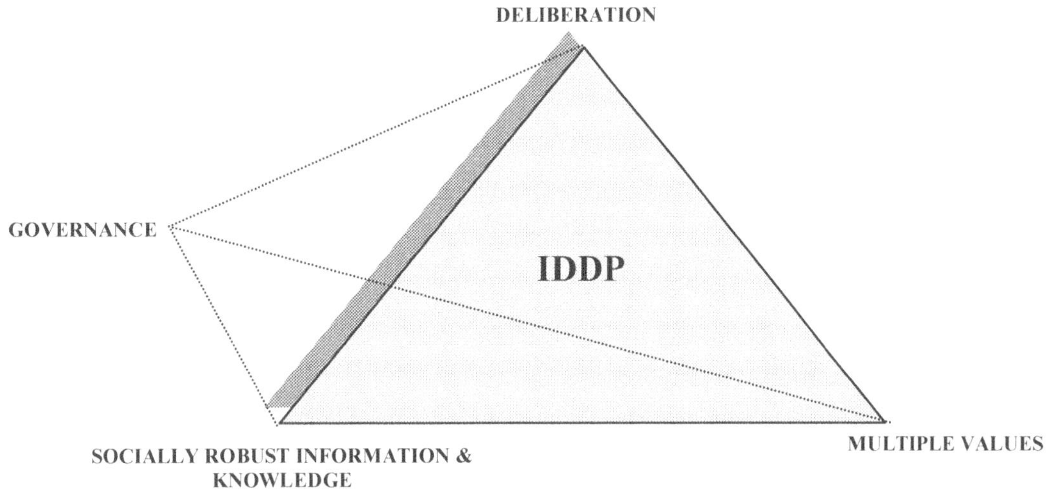

Figure 2.1 Dimensions of an IDDP

The features of a **deliberative decision process** are outlined in Box 2.1. The essence of a deliberative process is that decision (or at least, reaching a point upon which decisions can subsequently be taken) is the outcome of **interaction through dialogue** between a broad range of people with respect for their different values and viewpoints.

Several procedures, methods and tools exist now for deliberative decision-making. These can be distinguished in the procedural **platforms** upon which deliberation takes place (e.g. focus groups, workshops, panels, hearings) and the **tools** that help to facilitate the deliberation and the making of the decision (e.g. facilitation procedures, conflict resolution oriented facilitation, scenarios, visions, models, Geographical Information Systems, multi-criteria matrices).

Box 2.1 Features of deliberative decision-making (modified from Bloomfield et al., 2001)

(1) Social interaction takes place (through meetings, workshops, etc).

(2) Language-based discussion and debate.

(3) Facilitation of dialogue and debate, assisted by specific processes and tools.

(4) Respect of different values, viewpoints and initial positions.

(5) Design of process in order to exploit reflective capacity and allow participants to revaluate viewpoints and positions. Assumption that values and preferences can be transformed through dialogue.

(6) Participation and negotiation seen as having a value over and above the quality of the decisions that emerges.

The explicit reference to **integration** emphasises the need to see the deliberative process not alone, but in relation to the surrounding governance framework and in coordination with the management of information and the handling of multiple values and value systems (Figure 2.1). A second connotation of integration is that IDDP is not a one-off process or deliberative event, but a longer process consisting of several activities and events, and more importantly, embedded in the surrounding social and policy context.

By **socially robust knowledge**[3], it is meant knowledge that is fit for the purpose of the process, which cannot be described in limited terms as featuring scientific quality but must be fit for the context and relevant to the concerned social actors.

Information is a prerequisite for any type of decision and its management is crucial for the quality of the decision. Conventional water resource decision-making has emphasised the need for "hard", "objective" facts (as accredited by "experts" and the "scientific peer community"). Deliberative decision-making instead rests on a **"post-normal" scientific paradigm** in which an "extended peer community" becomes effective participant in the use of science for decisions (Funtowicz and Ravetz, 1990). This "extended peer community" does not consist merely of scientists and persons with some sort of institutional accreditation, but rather of all those with a desire and commitment to participate in the resolution of an issue. Scientific **quality assurance** depends on an open dialogue among all those affected, quality understood as a contextual property of scientific information. It may include formal or informal procedures, including definition of a set of quality criteria in **quality assurance protocols**. The "extended community" through appropriate process can then contribute to the **co-production** of socially robust knowledge.

Recognition of the **multiple values** present in any decision is a key characteristic of an IDDP. This distinguishes it from standard economic valuations which are based on the use of only one type of values (monetary). A deliberative decision process is designed instead, exactly in order to reveal value differences, make the participants aware on such differences, and hopefully pave in this way some shared ground for decision-making (see also Box 2.1, above).

However, it should be recognized that deliberative tools too and their design and conduct are not value free but they themselves pertain to certain "value systems". **Pluralism** in the use of tools is an advantage, as is pluralism within the application of each tool.

An IDDP takes place within a surrounding social, institutional and power nexus (**governance framework**). It is not possible for an IDDP alone to overcome the constraints and conditions set by this context. Power inequalities and exclusions in the social sphere will invariably lead to inequalities in the deliberative process unless explicitly mitigated. Participants with more resources, information or ability to express and articulate ideas will be privileged in the deliberative process. Hierarchical, centralised political or power structures will not be receptive to deliberative processes and decisions. Hence, an **inclusive, more egalitarian and deliberative governance framework** is necessary for a successful IDDP.

However, this relationship is not "top-down", but reciprocal. Deliberative processes can provide an impetus for changes in the governance framework and make it more inclusive and more deliberative itself.

The rules of the deliberative process should be clearly defined beforehand, preferably in an appropriate **regulatory framework** (supported by a conducive **constitution**). Rules should define clear procedures for identifying and

[3] After Gibbons, M. (1999) Science's new social contract with society. Nature 402: C81-C84.

including relevant 'stakeholders' with 'entitlement to participate', clear and verifiable lines of representation and definition of which level of involvement is foreseen for what kind of decision. Furthermore, regulatory change itself should be subject to deliberative mechanisms and processes.

WHY IDDPS FOR WATER RESOURCE PLANNING AND EVALUATIONS?

"In the sorts of issue-driven science relating to the protection of health and the environment, typically facts are uncertain, values in dispute, stakes high, and decisions urgent. [...] All too often, hard policy decisions have to be made where only scientific inputs are irremediably soft. The requirement for the "sound science" that is frequently invoked as necessary for rational policy decisions may affectively conceal value-loadings that determine research conclusions and policy recommendations" (Funtowicz and Ravetz, 1990).

There is a growing recognition of the need for deliberative processes in environmental decision-making. Several reasons have been given for this (Box 2.2).

The need for involvement of the public in decisions has been recognized in several recent international and European policy initiatives (Box 2.3). Likewise, water policy documents, including the WFD, have emphasised the requirement for the "active involvement" of the public in water resource management decisions (see Chapter 1).

Yet a review as part of the ADVISOR research of five water resource planning and project evaluation processes in Europe pre-WFD (see Box 2.4) revealed significant deficiencies in the actual practice of decision-making. The planning and evaluation processes followed were not explicitly defined and it was not clear when, how and by whom were decisions taken. Decisions lacked transparency and were characterized by a dubious use of science and appeals to expertise. Multiple values and perspectives were concealed under overriding, undisputable "strategic" goals. The result was ineffective decisions and social conflicts (Box 2.5).

Formulation of plans and evaluation of projects will be central to the implementation of the WFD, and more generally water resource management in Europe in the coming decades. Past experience raises questions on whether and how the quest for **integration** and **participation** will be realized in actual decision processes.

The dominant approach in current WFD guidance for implementation is to approach evaluation as an issue of **economic valuation**. However, novel research as part of the ADVISOR research project documented that economic values are only a subset of the several types of values that drive human behaviour (Box 2.6). This recognition of multiple values makes a strong case for an IDDP. IDDP provides a way to integrate different values and perspectives and to actively and directly involve interested parties in the making of decisions, allowing them to express and transform their values in the process. This holds considerable potential both in improving the quality of the decisions and in reducing conflict further down the implementation process.

Box 2.2 Some reasons for the growing interest in deliberative decision processes (after Pimbert and Wakeford, 2001)

Weaknesses of representative democracy
"Retreat of the state" to market forces has created a "governance void". Groups of people, especially the underprivileged, feel excluded from decisions that affect their lives. Distant governments′ decisions often fail to convey the wills of local peoples. Deliberative democracy is thought of as able to enhance democratic accountability, justice and empowerment of people.

Scientific uncertainty and complexity
Environmental problems (including water resources related) are very complex, with uncertain dynamics, outcomes and risks involved. Experts are seen as no better equipped to decide on questions of values, interests or acceptable levels of risk than any other group of citizens. Also some notable failures of scientists to foresee crises and a number of scientific controversies where experts came with very different opinions have reduced public trust in the supremacy of professional expertise and science. There is also a growing public suspicion of links between state and scientific expertise, pressing for the democratisation of decisions and the use of science.

Ineffectiveness in policy implementation
Centralized, top-down environmental policies often fail to deliver. Deliberative decision processes have the potential to improve the quality of decision-making by tapping on local knowledge and reducing conflicts at the design stage, increasing the likelihood that policy implementation will be more legitimate, effective, efficient and sustainable.

Conflicts
Exclusion of groups of the population from decisions, and especially those who face their consequences, leads to conflicts further down the implementation process. A deliberative process can ease these tensions by striving for consensual decisions at an early stage.

Box 2.3 Policy initiatives for public participation in decisions

White Paper on European Governance (COM 428 final, 25.07.2001)
With the underlying purpose of connecting more closely to its citizens while achieving more effective policies, the EU launched a wide debate on the reform of European governance with this paper in 2001. The foreseen changes in EU policy include: better involvement and more openness; better policies, regulations and delivery; acknowledgment of global governance and refocusing policies and institutions. Five principles for good governance underpin these changes: openness, coherence, accountability, effectiveness and participation.

Aarhus Convention, UN/ECE Convention on Access to Information, Public Participation in Decision-making and Access to Justice in Environmental Matters
This Convention was adopted at the Fourth Ministerial Conference "Environment for Europe" in Aarhus, Denmark, on 25 June 1998. Since then 40 countries and the European Community have signed it. This multilateral environmental agreement focuses on the interactions between the public and public authorities, providing a linkage between environmental and human rights. It covers obligations that Parties have to the public with respect to access to information, public participation and access to justice in environmental matters. The negotiation process of the Convention has build on the UN/ECE guidelines on "Access to Environmental Information and Public Participation in Environmental Decision-making" (1995), where public participation was recognized as one of the key elements for the long-term environmental programme for Europe.

EU Participation Directive (Directive 2003/35/EC of 26 May)
This Directive aims to contribute to the implementation of the obligations arising under the Aarhus Convention. It aligns Community environmental law with the second pillar of the Aarhus Convention (public participation in respect of the drawing up of plans and programmes related to the environment) and amends, with respect to public participation and access to justice, the Environmental Impact Assessment and the Integrated Pollution Prevention and Control Directives.

Box 2.4 Decision-making processes examined in Research Work Packages 1 and 2 of the ADVISOR project

Alqueva Multipurpose Dam in Portugal. A strategic reservoir at the Guadiana river formulated to meet the needs of the semi-arid Alentejo region in southern Portugal. The largest artificial lake in Europe (capacity over 4,000 hm^3, surface area 250 km^2, project cost around 1,800 million Euros). Part of an "Irrigation Plan" for the Alentejo Region.

Evinos dam in Greece. A medium size dam and reservoir (capacity 131 hm^3, surface area 3.5 km^2) in Western Greece designed to serve the water supply of the city of Athens. Part of "Athens Water Supply Plan 2020".

Grensmaas water project in river Meuse, the Netherlands. A multi-purpose river regulation project aiming to combine the commercial extraction of gravel (50 million tons/yr) with protection against flooding and development of a large natural area (1000 ha). Total cost of 250 million Euros.

NHP of Spain. Over 100 new dams and reservoirs and two major river diversions (including the river Ebro interbasin water transfer) aiming to re-distribute water resources from areas of surplus in north-western Spain to areas of deficit in south-eastern. The Ebro river transfer alone has a budget of 6 billion Euros.

Designation of a nitrate vulnerable zone in Ythan river, Scotland. Implementation of a special action programme for the control of nitrate pollution from agriculture in the River Ythan catchment, an intensively farmed lowland area in the North East of Scotland covering about 68,000 ha.

Box 2.5 Deficiencies in the five decision-making processes examined in Advisor

Lack of explicit procedure
Other than the Environmental Impact Assessment process, the evaluation of the examined projects and plans did not follow a formal, tractable procedure. There was no explicit, regulated procedure defining the steps that had to be followed for the formulation of a plan or the authorisation of a project. Scientific or economic evaluations were constrained to "rubber-stamping" pre-taken decisions or simply settling details of smaller significance (such as the reservoir's design in the case of Evinos).

Inappropriate treatment of scientific uncertainty and complexity
There was a lack of reliable scientific data as a basis for sound decisions. Complex information was treated in a subjective way, gaps in knowledge and uncertainties concealed under deterministic statements or calls for further research. For example, in the Ythan case the scientific understanding of the agro-ecosystem nutrient cycle was far from complete but once recommended for designation, the eutrophication of the Ythan estuary became framed as a problem related to agricultural nitrate pollution, despite high scientific uncertainty and different opinions between different government agencies.

Lack of public information and participation
The public and affected stakeholders had limited access to information and decisions. Their opinion was asked in ad hoc events (hearings) at a late stage and after main decisions had been framed. There was no explicit incorporation of the results of these hearings in the final decisions. Public input was constrained only to the discussion of the environmental impacts and not the overall rationale of the projects.

Neglect of multiple values and view points
Decisions were based on single, given and non-debated goals. For example, securing more water for economic growth was the unquestionable goal in the Alqueva, Evinos and Spanish NHP cases. In relation there was a limited consideration of alternative dimensions and selection criteria. Technical feasibility and one-off construction costs were the main factors taken into account in the Alqueva, Evinos and Spanish NHP cases. Environmental and social impacts were considered only in a reactive and remedial fashion through the EIA process, after key decisions were made. There was little concern for alternative viewpoints and alternative solutions. For example, the NHP of Spain did not consider in detail water demand management interventions as an alternative to new dams and transfers. In the Alqueva case, different designs of the dam were considered, but not alternatives to the dam that could also fulfill the goal of rural development in the Alentejo region. In fact, it seems that the decision for the specific projects preceded the formulation of goals and plans. The Alqueva and Evinos dams for example, were referred to many times in past decades as strategic reservoirs that had to be constructed. Plans were made to justify their construction, not to really compare them with alternatives.

Conflicts and ineffective decisions
Reactions from marginalized communities and stakeholders groups abounded in all five cases and impeded the implementation of the projects, rising also costs. In the Greensmaas case reactions from local inhabitants has blocked progress with the development of the project. Farmers have opposed intensively the designation of Ythan as a nitrates vulnerable zone. Demonstrations against the Spanish NHP in Spain and Brussels contributed to the repeal of the plan by the new government. In the Spanish NHP, it was documented that demand management could save an equal amount of water to this secured by the waterworks, at a significantly lower cost and with significantly lower environmental and social impacts. In the Evinos case, the additional quantity of water secured by the dam proved unnecessary after the end of the drought that prompted its construction. In Ythan, there are concerns that the expensive controls on farms in the designated zone will have only limited impact on the problem of euthrophication.

Box 2.6 Multiple values and motives: Lessons from valuing river restoration in Scotland

Economic valuations compare the costs and benefits of alternative options in monetary terms. The most popular proxy technique used for the monetisation of non-market environmental costs and benefits is the **contingent valuation method** (CVM) based on questionnaires that mainly elicit an individual's **willingness to pay** (WTP) for environmental improvements. Economic evaluation tends to assume all values can be expressed in a monetary equivalent. The behavioural model underlying this approach characterises a rational person as one concerned with selfish preference utility maximisation.

Research as part of the ADVISOR project revealed a much more complex picture regarding the behaviour of individuals and the factors that affect "how they value the environment". In order to assess the underlying motives for WTP of individuals a range of measures were developed and tested in a case study in Scotland. These measures included the categorisation of individual's primary ethical positions when answering a specific WTP question on the restoration of a river system. Some categories reflected animal rights while others were in accord with economic logic. The results show that, for those stating an ethical justification, **over 40% of individuals used non-economic reasoning to justify their responses to WTP questions**.

Further probing of motives for WTP was achieved in the case study by using measures based on a model of human behaviour from social psychology. These investigated the extent to which an individual takes account of attitudes, their perception of the extent to which they have control over an action, and the opinions of others (subjective norms). The action examined here was WTP to improve biodiversity in a given river system. The results show the model from social psychology far out performs the standard economic model used to explain WTP. This means the results from CVM have more to do with social psychological factors than economic ones and reflect a far wider range of concerns than individual self interest as defined in economic behavioural models.

In terms of using cost-benefit analysis to aid the WFD concern then arises because, for example, the assessment of alternatives via monetary costs loses too much information about how people make decisions between options. Cost-effectiveness alone is then inadequate for capturing the environmental values associated with water quality and quantity change. This makes the case for a **multiple use of evaluation techniques (eliciting different types of values).**

Further reading on the Theoretical Background of Deliberative Processes

Dryzek, J.S. (1990), Discursive Democracy: Politics, Policy and Political Science,Cambridge University Press, Cambridge.

Funtowicz S, and Ravetz J.R. (1990) Uncertainty and Quality in Science for Policy, Kluwer, Dordrecht.

Healey P, 1997 Collaborative Planning: Shaping Places in Fragmented Societies (Macmillan, Basingstoke, Hants).

Holmes, T. and Scoones, I, (2000), Participatory Environmental Policy Processes: Experiences from North and South, IDS Working Paper no 113, Institute for Development Studies.

Chapter 3

Designing and Running an IDDP

This section provides general guidance on **how to** design and run an Integrated Deliberative Decision Process for water resource planning and management. The guidance provided is **procedural** and describes the general tasks and steps for performing an IDDP (Figure 3.1). These are valid both for an IDDP for the preparation of an overall water resource plan as well as for the evaluation of individual projects, programmes or other water management decisions. Guidance on the use of specific deliberative tools is provided with more detail in Chapter 4.

> The process described in this section can be adapted to national, regional and/or local circumstances and the specific needs of the decision process. Political, organizational, cultural, hydrological and other environmental conditions should be taken into account when designing the deliberative process.

STEP 1: SET UP ORGANISATIONAL SCHEME

> Lack of **institutional responsibility** and undertaking of the IDDP will lead to failure. The IDDP should be embedded in a real decision-making process and should be initiated by those responsible for making and implementing the decision.

A basic first issue consists of defining **who** is the "**initiator**"/"**sponsor**" of the IDDP. These guidelines are written on the assumption that the authority/agency responsible for water resource planning or for a specific management decision (e.g. national environment agency, river basin authority, urban water utility, etc) should (voluntarily or as required by law, e.g. WFD) execute a deliberative decision process. Hence, they will be the initiators and main sponsors of the process and the event(s).

Reality however often defies this assumption; pressure for deliberation may come from the "bottom", i.e. from active citizen groups, non-governmental organizations, etc. Still, undertaking of responsibility (and not simply verbal support) by the decision-making authority is a prerequisite for the success of the process. IDDPs initiated by non-governmental or private organizations alone can be useful tools for active education / information, learning, research or demonstration purposes, but most possibly will have a limited decision impact.

> Organisers and sponsors yield power on the framing of the IDDP and hence the direction and emphasis of the procedure. Authorities/agencies are themselves "stakeholders" with specific interests in water resource planning and management. A river basin authority for example, may have an interest pro or against the development consent of a given controversial project.

It is advisable therefore that, although the initiation and sponsorship of the event may rest on an institutional actor, that a broader **forum** of stakeholders (with different viewpoints and interests on the issues) is formulated and takes responsibility for the organization and execution of the process. That is, **inclusion** and **participation** should be built in the process from the beginning.

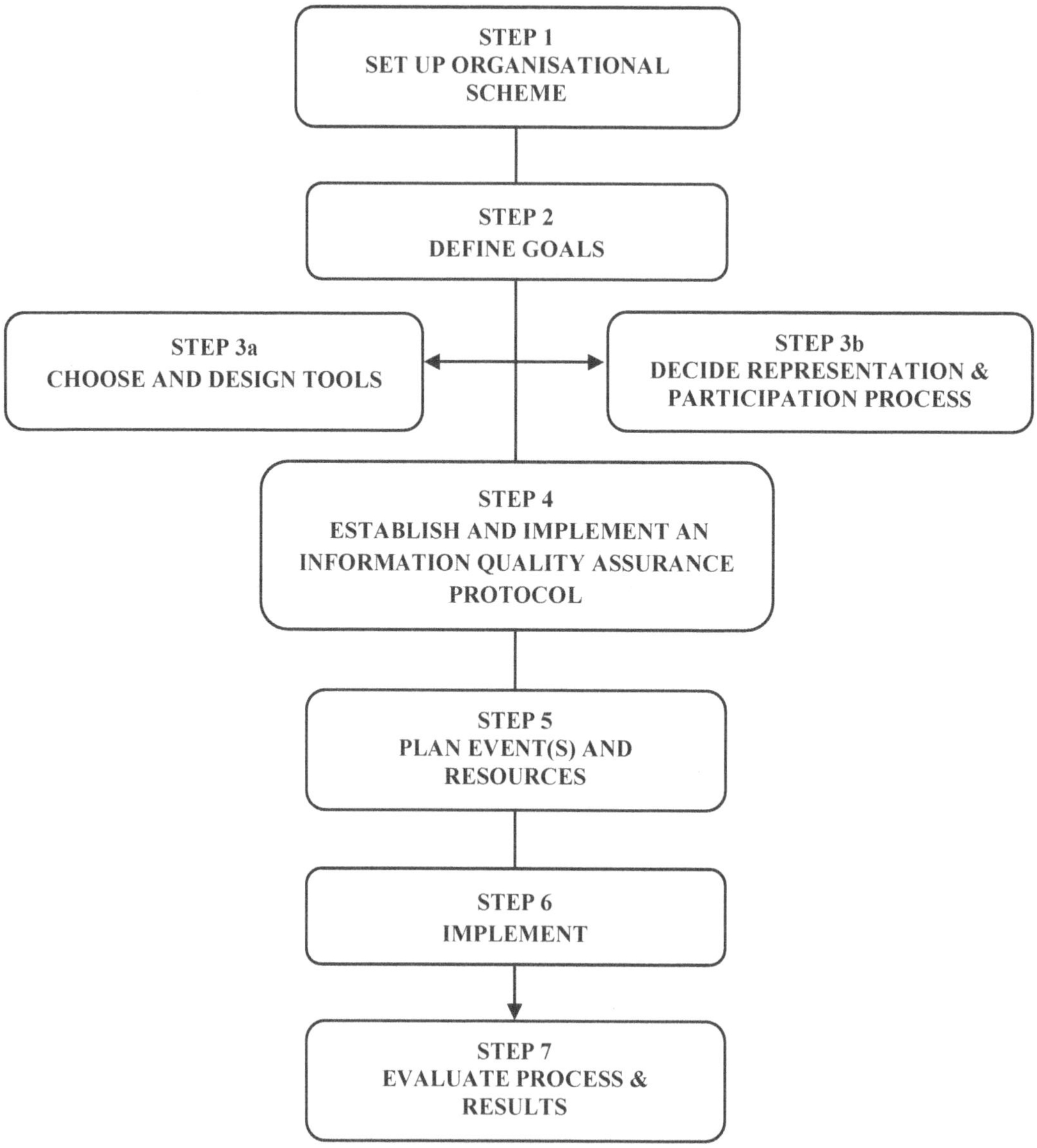

Figure 3.1 Steps of an IDDP

For example, river basin councils of stakeholders (if they exist), provide a good scheme for undertaking an IDDP of river basin planning under the overall responsibility of the river basin authority. For the authorization of a controversial water project, a forum consisting of representatives both from proponents and opponents of the work could be established. A forum for an IDDP for an urban water resource plan could include representatives from the urban water utility, regulators, customer groups and environmental NGOs. Sponsorship could then be divided between the members of the forum or public or private support sought as appropriate.

Forums should preferably have a **permanent structure and composition**, i.e. they should be parts of a broader planning and participation initiative (such as the WFD) rather than ad hoc groups formed for a specific IDDP or for an individual deliberative event.

In most occasions, the institutional actor or the forum will not have the needed expertise in order to design and run alone an IDDP. Professional support will be needed in the following tasks:

a) Implementation of IDDP and/or specific deliberative decision events;
b) Facilitation of deliberative discussions;
c) Information gathering, research (social and natural science) and analysis.

Some professional teams (research groups or consultancies) may possess the expertise to contribute jointly to more than one of the above tasks, providing thus for better coordination and economies of scope. The teams selected should be neutral, without a stake at the issues discussed and without ties to any of the interest groups.

Groups with relevant expertise can be found easily through an internet search or alternative open calls of tender or expressions of interest can be issued. Ideally, the professional teams employed should also have documented expertise in the above tasks. In some countries, there might be lack of local expertise (especially with respect to facilitation of deliberative processes or the organisation and management of deliberative tools). Language may put a barrier in contracting foreign experts. Training and capacity building in the use of the techniques may then have to precede the organisation of the IDDP.

A **Steering (planning) Group** consisting of core representatives from the forum together with the professional team should then be established. Its role will be to undertake the overall **organization** of the IDDP and the planning and execution of all practical tasks (steps 2-7, Figure 3.1). The division of responsibilities between the professional team, the responsible/sponsoring organization and representatives of the forum will depend on the local context. Generally, the professional team will take a leading role in management of the process where there is limited in-house expertise and previous experience with deliberative processes.

STEP 2: DEFINE GOALS

> Clarifying what is expected from the deliberation process and what goals will be served is a crucial initiation task. Creating over-expectations that are not fulfilled leads to lack of public trust in the process (and on future deliberative processes). Different parties may have different expectations from the deliberative process. Decision-makers may want a decision to be taken quickly, proponents of a controversial project to have opposition convinced, and marginalised actors to have time and space to express their point of view. It is the role of organisers to decide which goals are to be served, make them clear to participants and design the process accordingly. The prior establishment of a **partnership** and a **Steering Group** should serve for a broad agreement on the overall aims of the IDDP.

In the broadest sense the goal of the deliberative process will be to contribute to the making of a decision regarding a specific project or plan. There are different conceptions however of the kind of contribution that a deliberative process can offer (Box 3.1). Emphasis on different aspects merits different process designs.

Generally, the **goals** to be achieved through an IDDP may include one or more among the following:

- "Social learning", including:
  - Co-production of knowledge (*providing information to* / educating the participants and/or *receiving information from* the participants);
  - Revealing and appreciating different values and viewpoints;
- Conflict exposure or conflict resolution/consensus building;
- Revealing innovative solutions to problems;
- Identifying "win-win" options and forming partnerships around their implementation;
- Reaching consent on a common decision or set of actions (Action Plan).

Different goals relate to different planning stages and tasks (Table 3.1).

If the IDDP is applied to the evaluation of a specific water project or for judging on derogation from the environmental objectives of the WFD, then reaching a decision (or at least, a point upon which a subsequent decision can be taken) is the main deliberation goal. Ideally, individual decisions should be part of a longer planning process allowing hence for a longer and more comprehensive IDDP. As raised however in Chapter 1, often this is not the case.

The goals of the deliberative process should be positioned in relation to the broader policy process to which they contribute and aim to have an impact. **Linking the process to acute decisions** is a prerequisite if participants are to believe in the value of the exercise. It is important to distinguish at which level of impact the process aims at. If the aim is consultation, then clear rules on how the outcomes of the process will be taken-up should be stated clearly and transparently in advance.

Box 3.1 Conceptions of a deliberative process (after Holmes and Scoones, 2000)

Educative

Emphasis is on the deliberation process as a learning instrument with citizens. Goals range from simply providing participants with more information about an issue to empowering them to make collective political judgements and participate in future decision-making.

Consensual

Emphasises the search for a "common ground", i.e. recognition of different values and viewpoints and the agreement on shared goals and actions.

Instrumental

Emphasises effective end-results (decision to go ahead or not with a certain proposal, Action Plans, etc) directly applicable to the political or legislative processes.

Conflictual

Emphasises the need to give space to the expression of individual points of view (especially of those previously excluded) without constraints on discourse. Conflict and difference are prioritised over agreement and consensus. on the basis that thev can have an educative role and form the basis for future decisions.

Table 3.1 Provisional list of different deliberation goals in different stages of the planning process

Water Resource Planning Stage	WFD River Basin Planning Stage	Deliberation Model	Deliberation goals
Scoping	Management Issues		Co-production of knowledge
Goal setting	Environmental objectives	Educative Conflictual Consensual	Co-production of knowledge Conflict resolution/exposure
Alternatives Criteria	Gap Analysis & Identification of measures		Innovative solutions Different viewpoints
Evaluation	Evaluation of measures	Instrumental	Win-win partnerships Consent – Conflict resolution - Decision
Selection - Plan	Programme of measures	Instrumental, consensual	
Implementation	Implementation	Consensual	
Monitoring	Monitoring	Instrumental, consensual	Co-production of knowledge

A successful deliberative process will not be based on a one-off deliberative discussion event but on **a well-sequenced process of activities and events** (even if for the evaluation of a single project). It is generally advisable that each deliberative event avoids conflating many goals together. Each event should focus on a limited set of goals (e.g. education of participants in an early stage and dialogue and consensus later on) and altogether act to deliver the goals of the IDDP.

The deliverable of this first stage should be a **preliminary scoping document** setting down the organisation, responsibilities, general goals and expected outcomes of the IDDP. A **title** to the process or the individual event(s) has to be decided at this stage to facilitate communication to stakeholders and the public. The title could consist of a reference to the <aim/process of deliberation> and the <decision goal>. For example if the aim is to resolve conflict and deciding on a project, a title could be "Resolving our differences: deciding on the options for water supply". If the goal is to get to know each others opinions as a preparatory action for a water resource plan, an indicative title could be "Learning from each other: Preparing a Strategic Plan for the Sustainable Management of Water in …". The title given to the IDDP or a specific deliberative event is by itself a first, important framing of the goals and direction of the process. Care should be given not to constrain unwarrantly the deliberation options.

Framing the purpose and goals of a deliberative process is a most difficult and contentious task. For example, it is very different to frame a public debate about a proposed dam on the question of "having or not having the dam", "dam vs. other possible alternatives" (including demand management), or "which is the proper design of the dam in order to minimise impacts?" The way the question is framed is decisive for the debate that will take place and the outcome of the deliberation. The representative Steering Group should serve to safeguard an as representative as possible initial framing of the IDDP.

River basin planning for the WFD has pre-defined objectives, i.e. the achievement of the environmental objectives of the Directive. Note that this constrains the framing of the deliberative process (in comparison to a more open framing, e.g. deliberating generally on the future of water management in the river basin). In certain cases, local communities may not be aware and may not share the environmental objectives of the Directive or may have viewpoints on priority management issues that do not conform to the spirit of the WFD. Constraining public discussion on how to achieve the pre-defined goals of the WFD entails a trade-off with the deliberation principle of unconstrained discourse, and respect of all different viewpoints. Constrains not shared by the community may undermine broad public engagement in the process.

STEP 3A: CHOOSE AND DESIGN TOOLS

The next step is to choose the appropriate tool(s) for the achievement of the goals of the IDDP. There are several possible platforms for deliberation processes. These range from smaller (focus) **groups** of 6-10 people, to larger **workshops** of 20-100 participants or more. Work in workshops can be more manageable by dividing participants in smaller working groups of 6-10 people and then convening for synthesis in assembly.

There are several standard processes or techniques to organise and facilitate a deliberative exercise. The simplest form is a group meeting managed by standard **facilitation** techniques. This can be used for creative purposes (generating new ideas and solutions) as well as for mediation and negotiation and the building of sustainable agreements. In many situations, a group is manageable with maximum 8-10 people. Group facilitation builds on a two-tier logic (often corresponding to respective days of a 2-day meeting): the first is devoted to "**divergence**", where all ideas are freely debated opening up the spectrum of solutions. The second focuses on "**convergence**"; participants are helped with specific techniques (which may include voting) to conclude in a specific plan/decision. **Conflict resolution** processes usually include a group of representatives of conflicting interests coordinated by an experienced facilitator. The process follows a sequence of identifying the problem and relevant data, identifying alternative, innovative solutions that reduce conflict and then planning for implementation. **Consensus building** processes are somewhat similar, only the process follows a more open flow, with ideas and suggestions first listed (e.g. in a flip chart), discussed, voted and then debated in order to reach consensus.

The emphasis of these guidelines however is on the combination of deliberative processes with **decision-support systems** or **formal evaluation and planning methods**. The reader should refer to the literature for the full range of deliberative techniques available (see reading list below). Note that there is some confusion with terminology and practitioners in different disciplines or policy spheres often use different names for marginally different tools. Furthermore, there are very different variants of each tool, and open possibilities for modifications that can well expand the goals or range of applicability of each tool. There are also several possibilities for combining elements from several tools creating new "hybrid" tools (e.g. a citizens' jury for monetary valuation, a citizens' jury based on future visions, various combinations of focus groups with facilitated deliberation, groups or panels working on impact assessments, information technology based deliberations, etc).

Box 3.2 provides a checklist of questions that organisers should address when deciding which tool to use. It is not possible to provide here a full appraisal of the advantages, applicability and limitations of different tools. A careful literature research and study of applications can help addressing some of the questions of Box 3.2.

A toolkit that can address all basic water decision needs is provided in Chapter 4. This toolkit builds on three deliberative tools that were reviewed and tested in detail in the context of the ADVISOR project: **visioning workshops (VW), mediated modelling (MM) and social multi-criteria evaluation (SMCE)** (Table 3.2; details in Chapter 4). These three are complemented by tools for **Deliberative Monetary Valuations (DMVs)** and tools that allow for **broader community verdict** and opinion at the **final stages** of decision-making/planning, such as a **Citizens' Juries (CJ)** or **Consensus Conferences (CCs)**. These are also briefly reviewed in the toolkit section, although they were not explicitly tested in the ADVISOR project, and therefore the information provided rests on a literature review and not a detailed experimentation. These five tools can fulfil most basic requirements of an integrated water resource planning process in the context of the WFD. Furthermore, deliberations and decision-making as well as design of the tools and the process may be supported by a range of social science tools, such as **historical or stakeholder (institutional) analyses**, **surveys** and **value mappings** (see also Chapter 4).

Box 3.2 List of issues to consider when deciding which tool to use

1. **Conception and goals of deliberation**
   What is the main purpose of deliberation?
2. **Stage in the planning process and type of decision**
   Project (and type of), plan (and stage of), regulatory reform, policy intervention, or other?
3. **Type of representation sought**
   Open invitation, stakeholder based or random selection? (see more details below)
4. **Simplicity of deliberation vs. decision-aid capacity of the tool**
5. **Available time, financial and human resources vs. requirements of the tool**

Table 3.2 IDDP Tools in the ADVISOR Toolkit (bold denotes tools tested and developed as part of ADVISOR)

Tool	Platform	Short description
Visioning Workshop	Stakeholders Workshop (20-100 participants)	Two ½ day workshops evolving around the formulation of a shared desirable vision of the examined system in some time in the future and working to define the set of actions necessary to realize it.
Mediated Modelling	Stakeholders Workshop(s) (10-50 participants)	Participants develop a computer model during a series of workshops mediated by one facilitator, and then simulate alternative scenarios for management of the issues at stake.
Social Multi-criteria Evaluation	Stakeholders Focus Group + Interviews	Stakeholders define and evaluate list of alternatives and criteria and perform evaluation assisted by software matrix platforms. Emphasis on deliberation, conflict identification and resolution rather than axiomatic ranking of alternatives.
Deliberative Monetary Valuation	Focus Group + Population Surveys	Monetary valuation recognising multiple values in design of questionnaires and utilizing public input from focus groups meetings in design.
Citizens Juries/ Consensus Conferences	Randomly selected panel	Citizens' panel hears evidence on a controversial policy or scientific issue and reaches a verdict which includes recommendations.

Table 3.3 indicates advantages and disadvantages of each tool and the goals of deliberation for which it is most useful. Table 3.4 indicates which tools are more relevant for which types of decisions or stages in the planning process. An integrated deliberative decision process may **combine more than one tool** to achieve several goals in a longer-term planning process.

1. The choice of a tool frames the process. There is no neutral tool. Each tool builds on a specific mental model and **assumes a certain world-view**. Multi-criteria evaluation for example, assumes the ability to rationally distinguish between means and ends; alternative viewpoints might not share this initial framing/positioning. The hidden assumptions of each tool should be made explicit to participants and limitations recognized. A tool should not be "advertised" as a supreme instrument of solving problems, but only as a limited aid to debate. Complementary use of more than one tool should be pursued where possible and feasible.
2. The complexity or mental model of a tool may act to "estrange" some participants from the process. For example, participants who are less accustomed in thinking long-term and act on an experiential basis may feel alienated from the logic of a visioning workshop. Less educated participants, who are traditionally excluded from decision processes, may face problems in following the use of advanced decision-support tools. Organizers/facilitators should strive to explain the use of the tool and make sure that all participants follow its application.

Table 3.3 Appraisal and application of deliberative decision process tools (See Chapter 4 for details)

Tool	Advantages	Disadvantages	Deliberation goals
VW	Long-term thinking Educative Fosters dialogue and trust Creative thinking, innovative ideas "Win-win" partnerships	Too general, compromise statements may result Much dependant on quality of facilitation Limited time to assimilate information and reach concrete decisions	Social learning /Co-production of knowledge Innovative solutions Win-win partnerships Appreciating differences
MM	Creates a group learning process Educative Accommodation of participants with complexity Systems approach	Quantitative model building is over-dependant on professional support. Simulation model demands robust information.	Social learning / Co-production of knowledge Consensus building Innovative solutions Appreciating differences
SMCE	Quantitative and qualitative information used Use of formal assessment tools	Complicated, may constrain debate in comparison to more free-flow methods Over-dependant on professional support	Innovative solutions Win-win partnerships Appreciating differences Conflict resolution Decision
DMV	"Pragmatic", policy useful output (monetary value) Allows deliberation in the expression of monetary value	Constrained deliberation Difficulties in articulating it to the public Friction between deliberation and "market" logics	Decision
CJ/CC	Directly democratic – more legitimate representation Allows different perspectives/opinions to be appraised by citizens	Organisational/resource requirements Dependant on quality of panel Sensitive to framing	Social learning Decision

Table 3.4 Provisional allocation of tools for different planning and evaluation tasks (See Chapter 4) [4]

Water Resource Planning Stage	WFD River Basin Planning Stage	Project Authorisation or Derogation Evaluation	Tool
Scoping	Management Issues	-	VW, MM
Goal setting	Environmental objectives	-	VW, MM
Alternatives	Gap Analysis & Identification of measures	Alternative designs / solutions	VW, MM, SMCE, CJ/CC
Criteria		Criteria	SMCE (potentially VW, MM)
Evaluation	Evaluation of measures	Evaluation	DMV SMCE
Selection – Approval of Plan	Programme of measures	Final decision	CJ, CC

Further reading on deliberative decision process tools

COSLA (2002), *Focussing on Citizens: A Guide to Approaches and Methods*, COSLA, Edinburgh (available also at http://www.cosla.gov.uk).

European Environment Agency (2001), *Participatory Integrated Assessment Methods*, Technical Report 64, EEA, Copenhagen.

Guimarães Pereira, Â., Corral Quintana, S. (2002), From Technocratic to Participatory Decision Support Systems: Responding to the New Governance Initiatives, *Journal of Geographic Information and Decision Analysis*, vol.6, No.2, pp. 95-107.

Institute for Public Policy and Research (1999), *Models of Public Involvement*, IPPR: London.

Special Issue on PLA (Participatory Learning and Action) Notes, 40, February 2001, London: International Institute for Environment and Development.

See also articles on various other issues of PLA Notes at http://www.iied.org/sarl/pla_notes/backissues.html

[4] Planning implementation and monitoring tasks, following the final evaluation and selection of measures, are not addressed in detail in these guidelines.

STEP 3B: DECIDE REPRESENTATION & PARTICIPATION PROCESS

There is no such thing as the single best or *the* appropriate participant selection procedure. Each procedure has advantages, disadvantages and limitations. Decision on selection procedure and criteria should be **explicit**, **transparent** and **debated**. Limitations of representation should be recognised by organisers and the results of a deliberative process should not be branded as something more than what they really are, i.e. the results of a specific deliberation process and not necessarily the verdict of the whole community.

Selecting and recruiting the participants in the deliberative decision process is a key task. In an ideal inclusive deliberative process "all whose interests will be affected ought to have the opportunity to take part, and all citizens feel that their interests are being properly represented even if they do not choose to become involved themselves".[5] This is much easier said than done. In practice, there are four generic selection procedures and combinations of them (Box 3.3). Each selection procedure has advantages and disadvantages/limitations (Table 3.5). Choice should be made with respect to the goals of deliberation.

Note that the selection of tool and the representation procedure go hand in hand and should be considered together. Each of the tools of Table 3.2 is based on a different selection procedure. Visioning and mediated modelling workshops and multi-criteria evaluations work primarily with a stakeholder selection (of course if the goal is simply education and not contribution to decision making, then random, non-stakeholder selection would also do). Citizens' Juries / Consensus Conferences and Deliberative Monetary Valuations work with either random or representative criteria-based selection from the population.

Expect to face two key trade - offs when deciding which participants to select. The first trade-off is between size of the group and representative legitimacy of the process. Larger groups allow better representation but are counterproductive in terms of quality of the deliberation. Smaller groups deliberate better, but will tend to be less representative and hence the outcome of deliberation will have reduced legitimacy.

The second trade-off is between representative legitimacy and commitment in the process. The chances for a "smooth" dialogue increase by increasing the proportion in the group of those who are already experienced to deliberate or who are traditionally more active in community debates. Certain groups of the population might be more amenable to dialogue and agreement than others. Yet, the outcomes of such deliberation may be misleading in that they are not representative of real community constituency nor of real power dynamics that govern actual decision-making. On the other hand, in more representative forums (e.g. compulsory participation based on random selection), viewpoints may fail to be articulated and apathy may prevail.

[5] Bloomfield, D., Collins, K., Fry, C. and Muntun, R. (1998), "Deliberative and Inclusionary Processes: their Contribution to Environmental Governance", paper presented at the first ESRC "DIPs in environmental decision-making" Seminar, 17th December 1998.

Box 3.3 Participant selection procedures for a deliberative decision process (adapted from Holmes and Scoones, 2000)

Open invitation – "self-selection"

The event is advertised through mass media and individuals or groups can step forward and declare an interest to be included in the process.

Stakeholders´ identification and selection

The objective is to "get a representation of the whole system in the room". This means that in principle participants will not be able to appeal to somebody absent for the implementation of the decision. Stakeholders can be identified by analysts studying systematically the issue ("stakeholders analysis") and/or following suggestions from a first core of identified stakeholders (e.g. from the steering committee).

Criteria-based random selection

Random selection of a group from the population upon predefined representative criteria (e.g. gender, age, income group, profession, etc)

Random selection

Random (jury-like) selection of a group from the whole population.

Combinations

Combinations of the above generic selection procedures are possible. Open invitation for example may be screened upon certain criteria or be limited to individual representing constituencies (stakeholders). Stakeholder selection may be balanced for gender, age, inclusiveness or other criteria and a limited number of places may be maintained for self-invited participants on a first come-first serve basis.

Table 3.5 Comparison of different participant selection procedures

Selection Procedure	Advantages	Disadvantages	Deliberation goals
Open invitation	Unrestricted participation	Disproportionate representation – process "hijacked" by interested groups who show up or have resources, capacity Lack of democratic legitimacy Not applicable when restrictions in number of participants apply (as in more deliberative techniques)	Social learning /Co-production of knowledge (consultation) Appreciating differences
Stakeholders	"Those who matter and can implement" Invite the excluded	Limited democratic legitimacy Bias in selection Restrictions to people who might wish to participate Not representative of population Stakeholders might not always represent constituency (shifting positions in meeting may not reflect shifting positions of constituency) Tendency to employ established networks of actors	Social learning / Co-production of knowledge Conflict resolution Agreements – Action Plan "Win-win" partnerships Innovative solutions Appreciating differences
Representative - Random	Partly representative Higher democratic legitimacy	Bias / limits in representation criteria Limited room for representation in small group numbers Lack of power to implement decisions / non- representative of real decision making power	Decisions (verdicts)
Random			

Definition of geographical and temporal boundaries of decision is a difficult task that affects selection of participants. For example, a local (e.g. urban) water resource plan might affect populations in far distant regions (e.g. by water abstractions or new dams there). Including stakeholders from these regions in urban water resource planning may pose practical difficulties. Planning and decision-making at the river basin scale partly addresses these issues; still there will always be a need for local plans and decisions. Moreover, even in river basin plans there might be actors with important influence on the system from other policy spheres (e.g. from Ministries of Finance, from EU Directorates responsible for financing or environmental policy, global market forces such as international tourist operators, etc), that would arguably be difficult to involve in the deliberation process.

A **three-tier participation strategy** is recommended to account for the limited democratic legitimacy of deliberative processes (Figure 3.2). A **combination of tools and procedures** from these three circles should be pursued in an IDDP to maximise the democratic content of the process and the participatory coverage.

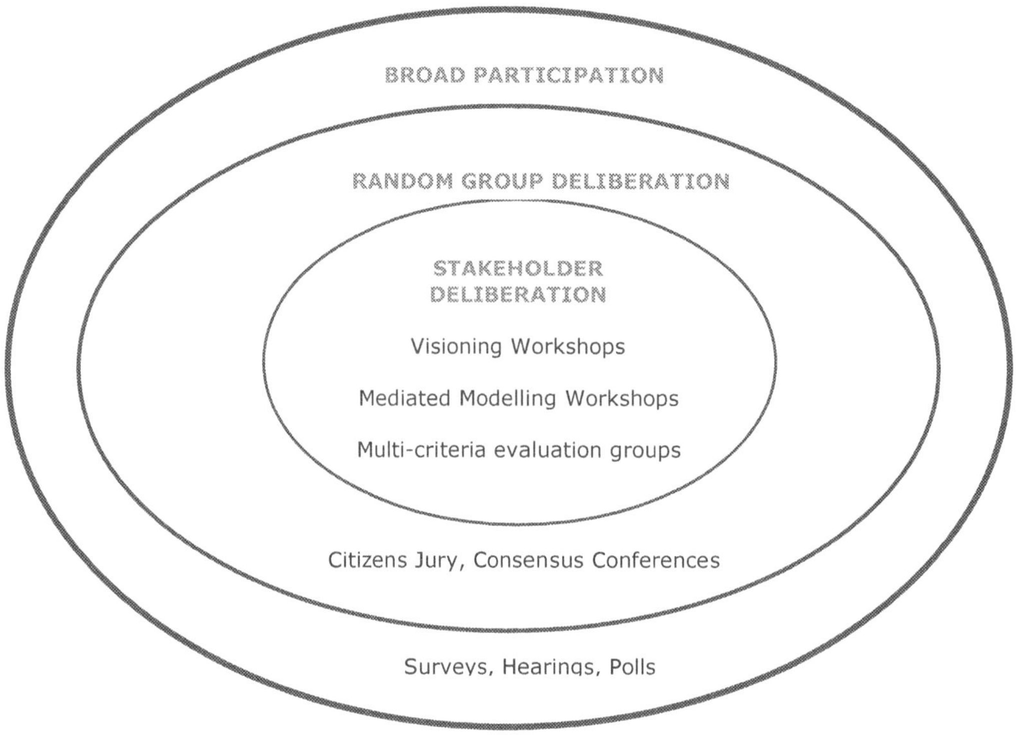

Figure 3.2 Three-tier participation strategy

The inner core includes deliberative processes with a selected number of stakeholders (such as visioning workshops, mediated modelling, multi-criteria evaluations). The permanent structure (council, committee, panel or other) established at the beginning of the IDDP provides the main platform for such events and should remain activated throughout the life cycle of the process. The middle tier includes additional deliberative processes (citizens' jury, consensus conferences, etc) that build on a random selection of participants and can serve to control democratically the decisions/proceedings of the inner core of stakeholders. A third "external" tier of control includes non-deliberative yet participatory processes (see Box 3.4) through which the activities of the stakeholders and the results of their deliberation are put for consideration to the broader community. The external circle can serve to inform the stakeholders on the issues that the community considers important and that should be addressed in their decisions (e.g. by public hearings or surveys) or to control the results of the deliberative process (e.g. by putting proposals subject to a referendum approval). Box 3.5 provides a relevant example from the United States.

Box 3.4 Platforms for broader participation

Opinion polls/surveys

Research questionnaires can be used to extract the opinion of the public in the debated issue and inform the deliberation and the decision proceedings. Care should be taken on the framing of the questions and the interpretation of results.

Public hearings

These are formal meetings with scheduled presentations offered. The process usually starts with the presentation of the full set of decision components to the public and the provision of a forum for answering all questions and collecting/defending opinions. For example, the results and rankings of a deliberative multi-criteria evaluation on alternative water supply solutions for a city can be presented and debated in one (or more) public hearing(s) open to all citizens. The authority responsible for the decision then can take both inputs of information into account when making and justifying the final selection.

Referenda

This is a more direct way of taking public opinion on a controversial issue (e.g. authorisation of a water project, approval of a controversial plan). It can only address simple "yes or no" types of question. Low turnouts and administrative costs are other drawbacks. Developments in "electronic democracy" (internet voting) may help to address some of these limitations.

No matter how well the design of the participatory process is, there will always be constraints relating to the surrounding social conditions and imbalances between participants in terms of power and resources (information, education, time and money). Power in real decision making will affect power in the deliberative process. Organisers/facilitators should be aware of such imbalances and try to manage them as best as possible. Support (financial, information provision) to underprivileged groups and structuring of the process so as to reduce imbalances should be pursued.

Box 3.5 The Blue Ribbon Committee *(after Wong, A.K. (1999), "Effective Public Participation in the Rate Setting Process: LADWP Blue Ribbon Committee on Rates", Gleick, P.H and A.K. Wong (eds), Sustainable Use of Water California Success Stories, Pacific Institute, Oakland)*

The Blue Ribbon Committee was appointed by the Mayor of Los Angeles in 1991 with the mandate to learn about water tariff design and supply and demand factors, decide on the principles to guide tariff design for the next decade and assist in the initial implementation of the tariff structure. In the **"inner core"** of the process was the committee composed of 12 citizen members with voting rights and 12 non-voting members from the municipal Department of Water and Power, the mayor's office and the city council. Citizen members were identified by the mayor's office as representatives of groups impacted by water prices and included homeowners, renters, landlords, the business community, the academic community, organised labour, developers, environmentalists, ethnic groups and neighbourhoods. Subcommittees reporting to the larger group for consideration were set up on "finance", "economic growth and development", "conservation and water recycling", "equity" and "public participation". A consultant responsible for collecting and analysing scientific information and facilitating deliberation in meetings and decisions coordinated the whole process. A Technical Advisory Panel was established to provide guidance on the consultant on how to explore different designs to water rates. The committee held over **75 meetings.** Some lasted a full working day, others well into the evening. The process started with an education of the participants on water rates and water management issues. Experts and representatives from other cities were invited to testify on rate structures.

External input and control to the process were provided through open **public hearings** (in each zone of the city) allowing for a broader participation of the interested public. Hearings were held at three stages: beginning of the process, after the first draft of recommendations and after the final proposed decision. Public hearings provided two-way feedback: from the public to the committee and from the committee informing the public and addressing concerns. Issues raised in the public hearings were reviewed and additional hearings held for confrontational issues. The Committee proposed an innovative, multi-functional and differentiated per use/zone water tariff structure that was adopted by the Mayor and implemented with successful results and lack of opposition.

Beyond selection, the **commitment** of participants in the process is another important challenge. **Public or stakeholder apathy** is often invoked as a reason against deliberative processes. The assumption of IDDP is that public apathy is not because of lack of interest, but because of lack of belief in that the process can make a difference. The only remedy is to ensure that the deliberative process affects real decisions. Public apathy can only be overcome by the building up of experience where participation really did make a difference to people's lives.

Lack of commitment however may also relate to lack of resources (time, money, etc). Economically underprivileged groups or special social groups (e.g. farmers, fishermen, single mothers) may face particular difficulties in committing to a long deliberative process or even a few days / full day event. In these situations, **compensation** for participants should be foreseen and included in the budget of the deliberative process. Conveniences should also be foreseen for participants with special needs (e.g. compensation to mothers for paying for children care, justification to employers for work leave).

Should participants be paid compensation? Arguments for compensation are based on the logic that participants are offering a public service and they should be rewarded for it or at least compensated for the time they devote on the process and the (real or hypothetical) income lost. Compensation is important to support low income participants and give them an opportunity to be included in the process. The counterargument is that deliberation should take place only where there is a real need to do so. Participants should be "willing to pay" for the costs of their participation and not be subsidised to do so. A forced deliberation may also turn out ineffective. Furthermore, one might argue that it is not proper to reward participants for exercising their democratic duties (for example, citizens are not paid to vote in the elections).

Lack of commitment may also stem from powerful **vested interests** that do not want to open up the debate and risk changes in the status quo. Successful deliberation erodes existing channels of influence and resistance should be expected from those who hold power. Powerful economic actors, politicians, etc, may not show up in planned events. In these cases, there is a danger that the deliberative processes delegate to debates between those excluded (or those who do not have power to implement decisions) with the risk of intensifying their feeling of impotence and disbelief in the ability of deliberative processes to deliver. Institutional commitment to the process from the beginning and clear linking of results to real decisions will help to commit powerful participants. Even if powerful actors show up, "strategic behaviour" and attempts to "control" or "sabotage" the process could arise. This highlights the need for embedding the deliberative process in a broader political and governance framework that is itself discursive and conducive to deliberation.

Further reading on issues of representation

Holmes, T. and Scoones, I, (2000), Participatory Environmental Policy Processes: Experiences from North and South, IDS Working Paper no 113, Institute for Development Studies.

O'Neill J, 2001, "Representing people, representing nature, representing the world" *Environment and Planning C: Government and Policy,* 19(4), 483-500.

STEP 4: ESTABLISH AND IMPLEMENT AN INFORMATION QUALITY ASSURANCE PROTOCOL

Proper management of the scientific information upon which decisions and deliberation are based is a major task of IDDP (see Chapter 2).

The requirement for "sound science" is frequently invoked as necessary for rational policy decisions. This may effectively conceal value-loadings that determine research conclusions and policy recommendations. A more relevant and robust guiding principle is **quality**, understood as a contextual property of scientific information (Funtowicz and Ravetz, 1990; Ravetz, 1996; Funtowicz, 2001).

Environmental *problematiques* entail multiple dimensions of analysis that cannot be *amalgamated* into a single scale of measurement because they pertain to different aspects and actors of the processes. Actors *talk* different *languages,* express uncertainty and concerns in different ways.

Social actors' contribution is not merely a matter of broader democratic participation, but a contribution to the co-production of knowledge. The procedure for assessing the quality of such policy processes is an **"extended peer review"**, performed by an extended peer community.

In the emerging context of **extended planning processes**, the exploration of the quality of socio-environmental planning procedures should be considered an essential element of the planning process itself. When the information used as input to policy making is complex, – deriving from several sources or evaluated through several criteria – uncertain and even disputed, there is a need for an agreed and credible set of procedures or guidelines by which the reliability of information can be explored and presented to all actors involved in the planning process.

For this purpose, a **Quality Assurance Protocol (QAP)** should be established at an early stage of the process providing normative guidance for the management of information in the decision process. At the end of the process, the same protocol can be used retrospectively in a descriptive way to assure the quality of information used during the process. Box 3.6 presents a provisional structure for the design of such a protocol. For a detailed description of the application of a QAP in the ADVISOR project see the project's Final Report (Chapter 10, available from http://ecoman.dcea.fct.unl.pt/projects/advisor). Box 3.7 illustrates the use of a QAP with an example from an application in a groundwater management case.

Satisfying the need for informed deliberation without giving a privileged status to **expertise** is a main challenge in the conduct of deliberative processes. Water decisions are **complex** and involve a degree of **technical and scientific knowledge** and capacity not available to ordinary citizens and stakeholders. Participants in deliberative processes may often feel ill equipped to discuss issues in detail. On the other hand, giving to scientific expertise a privileged status may disempower citizens, intensify feelings of ignorance and lead them to defer to perceived authority. Furthermore, it can give an illusive picture of "certainty" and knowledge where they typically do not exist.

Providing participants with information and training in key issues prior to deliberation might be necessary so that they all acquire a minimum acquaintance with the issues debated (e.g. establishing a long-term forum/stakeholder structure so that participants have more time and opportunities to deepen knowledge on issues). Different perspectives, methodological and data **controversies** and **scientific uncertainty** should not be hidden, but **transparently exposed** to the participants.

In many cases specific information and analyses might be needed for a specific deliberation and decision-making process. For example, in the case of the Blue Ribbon Committee (Box 14), studies were required to inform committee deliberations on different tariff designs and their impacts on specific income groups. Preparation of these studies might need additional research if primary data and related analyses are not available (for example, assessment of tariff impacts may require beforehand an analysis of water demand patterns and price/income elasticities of water demand for different groups). These studies and reports should be explicit about issues such as uncertainties, assumptions and reliability of information.

Further reading on quality assurance

Corral Quintana, S. (2000). "Una Metodología Integrada de Exploración de los Procesos de Elaboración de Politicas Públicas". Universidad de La Laguna, Tenerife. EC - Joint Research Centre: EUR 19724 ES.

Guimarães Pereira, Â., Blasques, J., Corral Quintana, S. & Funtowicz, S. (2003). TIDDDs - Tools To Inform Debates Dialogues & Deliberations. The GOUVERNe Project at the JRC. European Commission, Ispra: EUR 21880 EN.

Funtowicz, S. (2001). Peer review and quality control. In International Encyclopedia of the Social & Behavioral Sciences, ed N. J. Smelser & P. B. Baltes. pp. 11179-11183. Oxford: Pergamon.

Funtowicz S. and Ravetz J.R. (1990) Uncertainty and Quality in Science for Policy, Kluwer, Dordrecht.

Gibbons, M (1999) Science's new social contract with society. Nature 402: C81-C84

Guimarães Pereira, A. (2003) Quality of Information in River Basin Governance, Review of Case Studies: Information Vertex, Report Prepared for the ADVISOR Project

Box 3.6 A Quality Assurance Protocol (QAP)

The QAP summarized below was designed for the ADVISOR case study consisting of a Mediated Modelling Process in the Baixo Guadiana (see Section 4) and was divided into the following **six sections**: framing; social engagement; research team's role; research procedures undertaken; background information (scientific & technical information); communication of information. The number and order of these categories is based on the characteristics of the methodology applied in this case study, departing from the categories for quality assessment and information pedigree developed by Corral Quintana (2000). Hence, the organisation of the mediated modelling process has determined the structure of the protocol

Framing & social engagement

When looking into framing, one seeks to understand what was the basis to initiate a process of river basin governance in the chosen terms, i.e. how problems have been formulated taking into account the context in which they developed. Therefore issues such as historical and institutional basis; context description; stakeholders identification (who has initiated the debate); that can provide the justification basis for framing the problem in the socio-economic-political context, have to be looked at. Also, the context dynamics influence on information requirements and use will be analysed. The alternatives chosen to tackle the problem, their origin, or generation processes are also looked at to assess the quality of the problem formulation phase, as well as the assessment activities

Research team's role

Since it is recognised to experts their capability of translating the decision issue into alternatives and criteria it has also to be recognised that the planning process, and therefore its quality, may be influenced by that process of transformation. Quality Assurance categories are used in this case to assess that process, paying special attention:

a) To understand the *problematique* to be dealt with, and;

b) To transform information into alternatives, hypothesis and criteria.

Experts also choose the decision-aid tool which will be used to assess the different policy alternatives. It is clear that the election of one technique in detriment of other will influence the results of the analysis.

Two categories have been chosen to tackle the role of the analyst: Competence and Legitimacy. Competence is intended to assess the ability of the analyst to address a particular policy issue, in that sense three characteristics have been considered: the Experience of the experts in processing these problems; their *Adaptability* to deal with new components or characteristics of the issue and the way in which problems are *Structured.*

Research procedures undertaken

The quality assessment of the activities undertaken to evaluate river basin governance projects consist not only of the analysis of information used (including data sets, methodologies) through criteria of pedigree of information, but also the ways in which those that have performed the assessment have dealt with often observed information related problems (such as uncertainty, controversy – different schools of thought – data collection problems, knowledge gaps, generic models application, assumptions, etc.). Quality refers also to the transparency of the assessment procedures and to the robustness of the results that are made available through the studies. Robustness of results has to be assessed for their conclusiveness, their framing given the assumptions made at an earlier stage in the assessment, etc.

Background information (scientific & technical information)

One of the requirements demanded to any kind of software, instrument or object is its fitness for purpose to reach the aims for which it was designed or applied. Similarly the information or the knowledge used in the planning process must fulfil that requisite. This adjustment is analysed on the basis of three attributes: Adequacy, Accuracy and Completeness. Adequacy shall reflect the sensitivity of the information to reach the envisaged aims. Completeness is intended to examine whether the available data are complete or, on the contrary, they show some 'lacks' (i.e. incomplete series, stakeholder opinions not available...). At last Accuracy states the uncertainties related to the information analysed.

As important as the fitness for purpose is the *Applicability*; when the information cannot be used during decision-making processes, its quality will certainly be affected (independently of the degree of adjustment of the Information). Applicability is defined on the basis of *Accessibility* and of *Understanding*. Accessibility to the information and its sources plays an important role into the planning process - the information 'to be applicable has to be available'.

Box 3.6 A Quality Assurance Protocol (QAP) (continued)

The Information used in problem-solving processes should be *Reliable*, mainly when dealing with environmental planning, where stakes are high, decisions are urgent (and in many cases irreversible) and there is irreducible uncertainty and ignorance. The assessment of Reliability will be implemented through the use of two complementary characteristics: Control and Confidence. The former relates to both the description of the sources and the process of verification of the information while the confidence deals with the degree of legitimacy of the data.

Communication of information
Communication and reporting activities quality assessment will entail the assessment of information used for or produced from these activities. The criteria to evaluate information used for reporting and communication purposes will take into consideration not only information pedigree but also issues such as intelligibility (e.g. compliance with audience), style of reporting, management of uncertainties, etc.

The categories of a QAP roughly include the following topics:

- Fitness for purpose
  - (a) Correspondence between information used and problem – *Adequacy, Relevance*
  - (b) *Accuracy* of information
  - (c) *Comprehensiveness/completeness* of information
- Applicability
  - (a) *Access & Availability*
  - (b) *Intelligibility*
- Reliability: The evaluation of the Reliability of the information ranges from its internal aspects (such as its generation and verification) to those more external, such as those related to the acceptance of the specific information by the relevant community.
  - (a) *Control* (Sources, verification, colleague consensus)
  - (b) *Confidence* (Extended peer acceptance, legitimacy)

In relation to decision/deliberation tools, other categories can be considered such as:

- The transformation/encoding of information aspect in the Fitness for purpose category.
- Transparency
  - *(a) Methodology documentation*
  - *(b) Sources of information stated*
  - *(c) Arbitrariness - scientific set-up*
- Legitimacy
  Colleague consensus for methodologies used

The following attributes have been set up to explore the role of the analyst:.

- Competence
  - *(a) Experience*
  - *(b) Flexibility*
  - *(c) Problem-solving Methodology*
- Legitimacy
  - *(a) Verification*
  - *(b) Colleague consensus*
  - *(c) Extended Peer Acceptance*

Box 3.7 Co-production of a knowledge base as quality assurance procedure for water governance processes – the GOUVERNe process (Guimarães Pereira et al. 2003)

The *GOUVERNe process* can be divided into two major phases of quality assurance by extended peer review (Funtowicz & Ravetz, 1990; Funtowicz, 2001), one related to the whole process of unveiling, organising, representing and mediating knowledge; the other one related to the very involvement of the extended community into the process of dialogue, debate and deliberation.

As illustrated in the figure below, there is always a starting process of scrutiny of information. As this was a research project, initiated by researchers (and not by the relevant community[1]), the framing of the issues is certainly laden by research practice and perspectives. This is not to say that if the initiative had started within the community, there would be better accurate information, but probably the information would be more relevant for those concerned, yet probably not articulated in a way that researchers could use. However, by submitting the process from the very framing step to the community enables that a review of the issues is done on the basis of multiple perspectives, avoiding the pitfall of reductionism of complex situations. Indeed, while looking into the formulations of water issues, the most common aspects are related to scarcity and distribution in some areas and to the quality of the resource in others. Both descriptions are based on a *reductionistic* (economical and hydraulic) approach, which limits in many cases the issue to a supply side matter. This implies that the demand side issue is forgotten and so unsustainable planning policies and lifestyles are not usually under discussion.

A first formulation of the relevant issues by the researchers involved was necessary to help identifying who relevant actors are, what types of institutional contexts were researchers dealing with, what forms of community arrangements were there already that could be articulated into the process, etc. Through a preliminary social enquiry with identified actors (via institutional analysis or stakeholder mapping or any other methodology deemed appropriate), [experimented elsewhere, e.g. VALSE (http://alba.jrc.it/valse)] other actors could be identified and brought into the process.

Scrutiny of information from other than scientific or technical spheres, and indeed the whole quality assurance process, relies heavily in techniques of social research enquiry, such as in-depth interviews, surveys or even focus groups. In this endeavour for a co-production of socially relevant knowledge the main aim of the GOUVERNe process was precisely to explore ways of turning data, information, practice, *knowledges*, any relevant forms of understanding into "socially robust knowledge" (Gibbons, 1999). The challenges of this process were articulation and formulation.

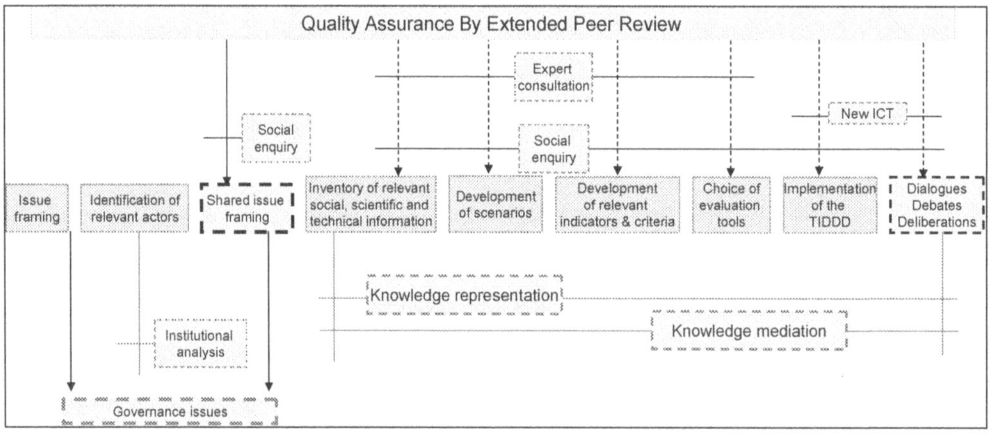

The GOUVERNe process including the challenges of knowledge representation and mediation in processes where plural views and different types of knowledge are implicated.

The GOUVERNe process was framed as a cycle of quality assurance; key aspects of quality assurance of this process are, on one hand ***fitness for purpose*** and on the other hand ***relevance*** of the information and tools used within all phases of the governance process, amongst others. The quality assurance process has addressed knowledge representation and knowledge mediation issues, including organisation, communication and exchange.

The final tuning contexts for the Hérault basin case study were organised according to focus groups. Three different groups were organised. A first group with researchers in relevant fields carried out in France, a second group with social actors of the Hérault basin, and a third group with social actors of the Köln basin in Germany. The quality of information and the relevance of governance issues were not a matter of discussion. This focussed on the option for river basin governance.

STEP 5: PLAN EVENT(S) AND RESOURCES

Having designed the sequence of the process, the tools to be applied, the events to be held and the participants to be involved and having secured and planned for the provision and management of information, the next step is to **plan the actual execution** of the event(s) (e.g. workshops, group meetings) and accompanying activities (for instance, research and/or surveys).

Time and financial resources are the main constraints for an effective IDDP. Water management decisions have typically short time frames. Budgets dedicated to deliberative processes are often low. Trade-offs between duration/cost *vs.* quality of the deliberative process will have to be made. The deliberative process however should not be compromised below certain quality minima for the sake of a fast decision or a limited budget. It is "better" to opt out from doing a deliberative process if this is not feasible, rather than doing a bad one and damaging trust in future deliberative processes.

After settling the main issues, the Steering Group should formulate and agree on a detailed **work-plan** and **timetable** of activities. This should include both **preparatory actions** (research, contact with stakeholders, provision of information on the process, awareness of the local population on the initiative) and the actual implementation events and actions (workshops, group meetings, surveys, etc). There might be scope to produce first a **general work plan** where specific events are described only in general terms and then, in due time but well in advance of events, elaborate on **specific work plan** for each event/task.

The detailed work programmes should be written down to the minute detail providing specific guidance on the members of the organising team (if necessary, for each responsibility separately) on how to manage the process. For example, a work programme prepared for the running of a scenario workshop as part of ADVISOR project research (See Chapter 4) included separate guidance for facilitators, secretariat, and assistants for each hour session of the workshop, and was supported with key reminders (from theoretical or local specific issues that might be encountered to practical details, such as having brought the proper pens with).

Budgeting of the process is a key management task. A **budget plan** should be prepared at the beginning of the process. Box 3.8 presents a provisional list of the main cost categories. An indicative budget of an IDDP cannot be given in general terms; much will depend on the level of commitment and detail put in the process. For example, the Blue Ribbon Committee (Box 3.5) held 75 meetings plus several public hearings and a number of research studies were carried out to support its activities. Evidently, such a high level of commitment increases the chances of obtaining meaningful results but also increases budget. In certain cases, a more targeted and less demanding IDDP may well suffice (e.g. running a citizens´ jury for a confrontational plan or doing a modelling workshop to identify the important issues in a river basin). However, the rule of thumb is that the more committed, comprehensive and "integrated" the deliberative process is the more the chances of success increase in the longer term.

Box 3.8 Main cost categories for an IDDP

Research and Information Support

This includes cost for the collection of data and the preparation of support studies. It might be delegated as a package to a research group or a consultancy. Costs will depend on existing availability of data and research/analysis requirements.

Organisation / Facilitation Support

This includes personnel and office costs for the preparation and management of the process and for expert / facilitation services in the use of tools and the execution of events. Costs will depend on the decree of "in-house" expertise and support capacity, requirements of the tool used (e.g. software related costs, application support costs), quality/cost of facilitator(s) contracted, number of facilitators required, etc.

Meetings

These include costs for renting a place for meeting(s), workshop(s), etc (if not in existing premises of the organisers), consumables for the meeting (stationery, etc) and provision of coffee/lunch/dinner to participants (if deemed necessary).

Compensations

In the case of work leave, these might have to be determined on the basis of the participants' salary or a proxy for income lost. For commitment in evenings or weekends, special arrangements should be sought as applicable and acceptable in each case.

Dissemination

These include costs of producing and where relevant posting information leaflets and publications, and costs of conducting and preparing press conference(s).

STEP 6: IMPLEMENT

Specific guidance on implementation cannot be given, as the details will depend on the tools employed. Section 4 provides more detail on the implementation of specific tools. Some more or less common preparatory and support actions will be needed however whatever the tool(s) used (Box 3.9).

In certain cases, deliberation may need to be supported by accompanying **social science research**. **Interviews** with stakeholders for example may help to identify power dynamics and contribute to the structuring of participant selection. **Surveys** may be used to elicit community opinion and inform the debates. The application of social multi-criteria evaluation in the Spanish case study of ADVISOR made such a combined use of deliberation with different social science research tools (See Chapter 4).

Surveys of community values or behaviour drivers may also be carried out early in the process (by professional social scientists) and help organisers in the design of the process and the choice of tools. For example, a river basin management authority may wish to know the extent to which the public regards a change (e.g. a river restoration) as a matter of market or non-market concerns (and hence decide the extent to which the decision will be "political", deliberative or based on an economic valuation). Alternatively, value/behaviour mapping may be used in order to identify which dimensions/criteria the public considers as important and hence provide likely "weights" of such factors for a multi-criteria assessment. Box 3.10 gives an example of the application of a behaviour survey as part of the ADVISOR project.

Box 3.9 Main preparatory and support tasks

Public Awareness

Awareness of the public in the area where the IDDP takes place is a prerequisite for its success. Relevant activities include dissemination of information material (leaflets, brochures, etc), advertisements (public kiosks, billboards, posters, advertisements in mass media), articles in the local press and interviews in broadcasts. A constant flow of information and updating on the progress of the process should be maintained. Awareness activities will benefit if IDDP is embedded in a broader policy initiative (e.g. a plan).

Engaging participants

Effective channels of contact between organisers and participants are important. Involving the participants from the beginning (through a permanent partnership structure and a steering committee) is vital. Supports may be needed to maintain contact, such as publishing a regular brief or newsletter on the activities of the committee. Official letters of invitations to events, information packages on the events or the tool, etc, can all help. Other supports, such as workbooks, may also be helpful in reporting progress to participants throughout the events organized under the deliberative process.

Practical details for meetings

There are several practical issues, which tend to be underestimated but prove decisive for the success or failure of a deliberative meeting. These include selection of the location for the meeting, characteristics of the premises (light, space, etc), distribution of activities in the space, securing availability of materials, timing of intervals, quality of catering, etc. An interesting suggestion is to hold meetings in far away locations (e.g. in a hotel resort), distancing participants from everyday environment and allowing them to build more direct relationships and get to know each other better.

Training

Members of the team supporting the conduct of the deliberative events should be well experienced in the use of the tools. If this is not the case, training must be provided well in advance. If members of the team are not well acquainted with the tool and processes used it is suggested to run first a **rehearsal** (test event) with hypothetical participants playing roles.

Dissemination

It is important that the results of the IDDP and of individual events receive broad coverage. A report summarising the proceedings main results should be available in due time after an event. Press conferences should be built in the events (e.g. embedded in the last day and conclusion of the programme).

Box 3.10 Valuing mapping and motives in the Tummel catchment Scotland

A range of survey/interview techniques can be used to obtain quantitative and qualitative indicators to map out the underlying motives behind public framing of water use issues. Scales used to measure **attitudes**, **norms**, **behavioural control** and **ethics** can be employed as independent indicators or combined in an overall assessment model. This provides the opportunity for river basin managers to probe multiple motives lying behind public behavioural intentions (e.g. WTP/WTA for river basin schemes and management options).

An example is the ADVISOR Tummel catchment case study where **multiple values informed public WTP** for changing hydro dam river flows to bring benefits for biodiversity. One measure of specific attitude towards payment for improving biodiversity was achieved by combining two components to create an attitude scale: first a given belief about a behaviour 'B'; and second the individual's evaluation of that belief 'E'. For illustration, we provide below a set of questions used to elicit B and E respectively:

"Paying more for electricity to restore biodiversity will increase the diversity and abundance of plant and animal species in the Tummel area.
Extremely likely 1 2 3 4 5 6 7 Extremely unlikely"

"Increasing the diversity and abundance of plant and animal species in the Tummel area is:
Extremely bad 1 2 3 4 5 6 7 Extremely good"

The sum of the product of the responses to several such questions framed on a scale of 1 to 7 provided a quantitative measure of public attitudes towards payment. Similarly, quantitative measures of subjective norms and ethical positions were created. Socio-economic data collected at the same time can be used to analyse group specific clustering of attitudes, norms and ethical beliefs to inform the IDDP framing, sample selection and analysis.

Value mapping allows a river basin manager to discover whether a proportion of the public regard environmental change as a matter of non-market concerns which relate to moral issues, their community, constraints on their behavioural intentions and specific or general attitudes. As a result they can determine the importance of using different approaches to participation and the likely weight of such factors in a multi-criteria assessment. A mapping out of underlying motives and different values can aid the framing of IDDP. More information on this case study is available from http://ecoman.dcea.fct.unl.pt/projects/advisor/iddp/

Facilitation (i.e. management of the deliberative process) is a crucial task. It is recommendable that facilitation is assigned to a neutral and experienced facilitator with no ties to the organisers and preferably, an outsider to the "system". One or more facilitators may be needed depending on the nature of the process. Allowing groups to self-organise can reduce the number of facilitators needed. The facilitator(s) should have a helping attitude, while conducting the events with authority and integrity.

There are different perspectives and approaches to facilitation. For some, a good facilitation should have minimum interference in the process and the debate and should be constrained on maintaining procedural and methodological rules and the "flow of the process". Others advocate a more active role for facilitators in order to "support" participants with expression difficulties, lack of self-assurance, etc, to be included in the debate. The counterargument is that it is impossible to account for all personality differences in a group and that it is better to let debate evolve with minimum interference, reflecting also real life dynamics. Clearly, there is not one *single best* facilitation approach; rather different approaches for different goals and for different group dynamics.

Facilitation techniques can reduce but cannot eliminate a certain degree of bias. They are constrained by the initial framing of the issue, goals of the process and features of the tool used. Facilitators are humans and cannot avoid managing the process according to their own mental models. For example, deciding whether to put emphasis on the inclusion of the excluded or not, is partly a methodological issue, but reflects also broader "viewpoints" about the scope of deliberation. Furthermore facilitators have a "stake" in the process, in terms of seeing it succeeding. Hence they may be tempted to promote dialogue, consensus and decisions even if conditions are not mature. The more experienced the facilitators the more aware they will be of the inherent limitations of the facilitation process.

Further reading on facilitation

Kaner, S., Lind, L., Toldi, C., Fisk, S. and Berger, D. (1996), *Facilitator's Guide to Participatory Decision-Making*, New Society Publishers, Philadelphia.
Hart, L.B. (1992) *Faultless Facilitation*, London, Kogan Page.
Howell, Johnna L. (1995) *Tools for Facilitating Team Meetings*, Seattle, Integrity Publishing.
Heron, J. (1993) *Group Facilitation: Theories and Models for Practice*. London, Kogan Page.
Rees, F. (1991) *How to Lead Work Teams: Facilitation Skills*, San Diego, Pfeiffer & Co.

> Resolving differences in group discussions and reaching consensus (if needed) is a particularly difficult task. Different tools embed different approaches to agreement and conflict resolution. Additional facilitation may be necessary in order to resolve differences when joint agreements are needed (e.g. in statements, valuations, etc) but differences persist. Voting is a possible instrument, but will be of limited legitimacy if participants feel that the distribution of voting rights is not proper. Maintaining minority positions together with majority ones is another option but may not be feasible in all circumstances. Other well disseminated methods include the *Nominal Group Technique* or *Delphi* Processes.

STEP 7: EVALUATE PROCESS AND RESULTS

Evaluation of the deliberative process as well as individual events is crucial in order to assess impact and added value and learn in order to improve future processes.

A distinction needs to be made between the evaluation of the achievement of **outcomes** and the evaluation of the **procedure** (although this distinction is not always easy to maintain as key outcomes of deliberation may relate to procedure *per se*). By outcome we refer to concrete actions such as educating the participants, producing an Action Plan, supporting a decision, formulating an action partnership, resolving conflict in a controversial issue, etc. As explained in step 2, each deliberative process might aim to emphasise different goals; evaluation should then be tailored to the goals of the specific process. For some goals, evidence of concrete follow-up action (e.g. implementation of a joint project, agreement on an Action Plan, formulation of a partnership, etc) is a main indicator of success. However, much care is needed when attributing to a deliberative process a positive policy outcome. Several other factors may have contributed to a positive policy outcome other than deliberation. On the other hand, deliberation may have also had considerable "side effects" and positive contributions going unnoticed.

In evaluating the substantive outcomes of a deliberative process from an instrumental viewpoint, a key issue is whether deliberation led to concrete, implemented changes. A main critique against deliberation is that the apparent achievement of procedural goals at the end of the process (e.g. conflict resolution and agreement on an Action Plan), often is not followed by real change in practice (i.e. conflicts and differences re-emerge at the implementation stage). A possible explanation is that the deliberative process may not be sufficiently embedded in real decision and power structures (see Chapter 2). ADVISOR project research on citizens' behaviour towards biodiversity benefits from river restoration in a Scottish catchment highlighted yet another important factor. Responses change considerably when from general commitments to a goal (e.g. biodiversity restoration) citizens are asked to resume responsibility for this change (e.g. in terms of declaring willingness to pay for it).

For the appraisal of procedural outcomes such as the education of participants, the co-production of new knowledge or the resolution of differences, a different approach needs to be followed since results are not directly evident. An **evaluation questionnaire** can assist in this task.

> In designing evaluation questionnaires it is preferable to devise indirect questions and reveal changes rather than asking participants directly their opinion after the process. For example, if the aim is to assess whether knowledge of the participants on certain issues has increased or not, an open question about the nature of water problems could be posed before and after the IDDP and quality of the responses compared. This would be preferable than for example asking participants directly at the end of the process whether they feel they have more knowledge than before. The first approach is more objective and reduces any possible bias in participants' responses (e.g. due to lack of understanding of the questionnaire, a "feel good factor" after the event or a wish to satisfy organisers).

Procedural goals of an IDDP include deliberation/participation, addressing multiple values/viewpoints and qualitative use of information (see Chapter 2). Box 3.11 provides a provisional list of questions by which to assess the quality of a deliberative process. The quality of information deployed throughout the IDDP may also be appraised at this stage through the use of the QAP (see Step 4 and Box 3.6).

The specific questions posed will depend on the specific goals of an IDDP. Appraisal should be performed mainly by the organisers and the Steering Group. Questionnaires for the participants and other proxy indicators may have to be devised for a more objective appraisal of certain criteria.

Box 3.11 Indicative questions for evaluating quality of deliberative process
(Modified from IEMA (2002), *Guidelines on Participation in Environmental Decision-Making*, Perspectives Series. Lincoln: Institute of Environmental Management and Assessment)

Was the process inclusive?
- Did the process include a wide range of participants, representative of interested and affected parties and individuals?
- Were adequate resources provided to allow a range of individuals to participate?
- Were meetings and events arranged at times and venues most convenient for participants?

Was the means of communication appropriate for all interested and affected participants?
- Was non-technical language used?
- Was information translated into appropriate languages and transcribed into appropriate formats?
- Was every participant provided with equal access to documents, sources of information etc?

Was the process open and transparent?
- Were the aims and objectives of the participation programme clear and agreed with all participants?
- Were the expected inputs to and outputs from the process clear?
- Were the limitations on the participation programme made clear?
- Was the process and its outcome well published?
- Was the agenda for discussion agreed with the participants?

Was the process interactive?
- Did techniques for deliberation allow all participants to both contribute and receive information?
- Was group discussion and debate promoted?
- Were techniques used that allowed participants time to understand documents and information enabling them to take part in the discussion more effectively?
- Was there sufficient time for discussion of issues and for responses to be made?

Was the deliberation process relevant to the decision-making process?
- Was the mechanism for deliberation relevant to the objectives of all participants?
- Was all the information provided relevant, accurate and up to date?

Was the process credible?
- Were values and concerns elicited as well as information being provided?
- Was the deliberation process constructive?
- Were opportunities provided for disagreements to be aired and resolved as far as possible?
- Was evidence provided of how concerns have been dealt with?
- Was peer review and independent verification of data and knowledge promoted?
- Were all the participants given an equal opportunity to voice their concerns?
- Were all issues of genuine concern responded to?
- Were participants satisfied with the way the results were used in the decision-making process?
- Did deliberation inform and influence the proposed activity and decision-making process?

Box 3.12 recaps on the main practical tasks of the organizers of an IDDP have to take care of. Box 3.13 reminds the critical issues raised and some of the potential responses in order to address them.

Box 3.12 Running an IDDP: the organisers' "to-do list"

- Collect all the relevant background information to initiate the process
- Set up a Forum of stakeholders (or, preferably, embed IDDP in existing Forum)
- Ensure the support of a professional group(s) to assist in research, deliberation tool and facilitation
- Set up an IDDP Steering Group
- Prepare a scoping document with title and expected outcomes of the process
- Prepare a detailed work plan with an accompanying time table presenting activities to be carried out, events to be held and participants to be involved
- Prepare a detailed budget for the process indicating allocation per category of costs and sources of funding
- Establish a Quality Assurance Protocol
- Prepare specific work plans for deliberative events to be carried out
- Carry out information and training sessions for participants
- Execute deliberative events
- Pose results to broader public scrutiny (hearings, referendum, etc) (if relevant)
- Prepare reports with proceedings of the events and the whole process and evaluation of results

Box 3.13 Problematic Issues in an IDDP and some useful responses

Framing of goals and process => *Deliberate goals early on in Steering Group*

Putting or not boundaries/constraints in focus of deliberation? => *make sure that goals are shared by Steering Group / Forum*

Partiality of organisers/sponsors => *ensure a broader organisational partnership/ scheme*

Partiality of managers of the process, scientists, etc => *make it explicit and transparent*

Lack of **democratic legitimisation** => *use more than one tool and representation/ participation processes*

Public apathy => *link to acute decisions*

Vested interests control => *change governance framework*

Limitations of tools/dependence of outcomes on design => *use more than one tool in combination*

Power imbalances => *manage process in a transparent and inclusive way. Support weaker participants. Embed in broader governance framework.*

Scientific complexity => *seek assistance but maintain a Quality Assurance Protocol*

Limited technical knowledge of participants => *support with information sessions*

Time and resource constraints => *ensure that adequate time and financial resources to achieve deliberation goals are available before committing to process. Design process and expectations appropriately*

Facilitation *of deliberation => hire professional facilitator. Recognise and make explicit to the participants the limitations of facilitation*

Chapter 4

An IDDP Toolkit

This section provides more details on the application of specific deliberative decision support tools for water resource planning and evaluation in the context of the WFD. The list of tools presented covers only a small portion of the options available to the organizers of an IDDP. Still, this selection of tools - "**toolkit**" - suffices to cover all basic needs of water resource planning and decision making.

The toolkit builds on the three tools developed and tested in the ADVISOR project: Visioning Workshops (**VW**), Mediated Modelling (**MM**), Social Multi-criteria Evaluation (**SMCE**) (Box 4.1). Some decision needs however, such as getting a monetary value about different alternatives or getting a representative community opinion/verdict on a controversial issue, cannot be met only by these tools. The toolkit is thus complemented with two additional tools that were examined and reviewed, but not tested in the ADVISOR project: Deliberative Monetary Valuations (DMV) and Citizens' Juries (CJ) / Consensus Conferences (CCs).

Box 4.1 Deliberative tools and case study applications in the ADVISOR project (Final report available at http://ecoman.dcea.fct.unl.pt/projects/advisor)

Visioning Workshop for the Sustainable Management of Water in the island of **Naxos** in **Greece**.
Mediated Modelling to Support Public and Stakeholder Participation in Water Resources Planning and Management: The **Baixo Guadiana** Experience, **Portugal**
Social Multi-criteria Evaluation of water management alternatives for improving water supply in **Costa del Sol Occidental, Malaga, Spain**.

In principle, each deliberative tool is designed to be applicable for the whole decision making and planning process, from scoping and framing to evaluation and selection. However, testing and scrutiny in the ADVISOR project revealed that the strengths and weaknesses of the tools make them less or more applicable for different stages of the planning and decision process.

In particular, VW and MM are ideally suited for an earlier **Decision Framing phase** paving the way to a future decision, whereas SMCE, DMV and CJ/CCs are more helpful in the actual **Decision Making phase** (although they can also contribute to framing) (Figure 4.1). There are various possibilities for modifications in the design of the tools, or combinations with other tools, that could expand their scope and usefulness. Alternatively, more than one tool could be combined as part of a longer planning or decision process.

Use of tools such as SMCE, DMV or CJ for decision-making, does not necessarily imply that they can lead *alone* to actual decisions. The tools are *aids* to decision making. Inputs from more than one source and tool should be used by responsible authorities (including consultation with the broader public; see Section 3) for reaching a final, justifiable decision.

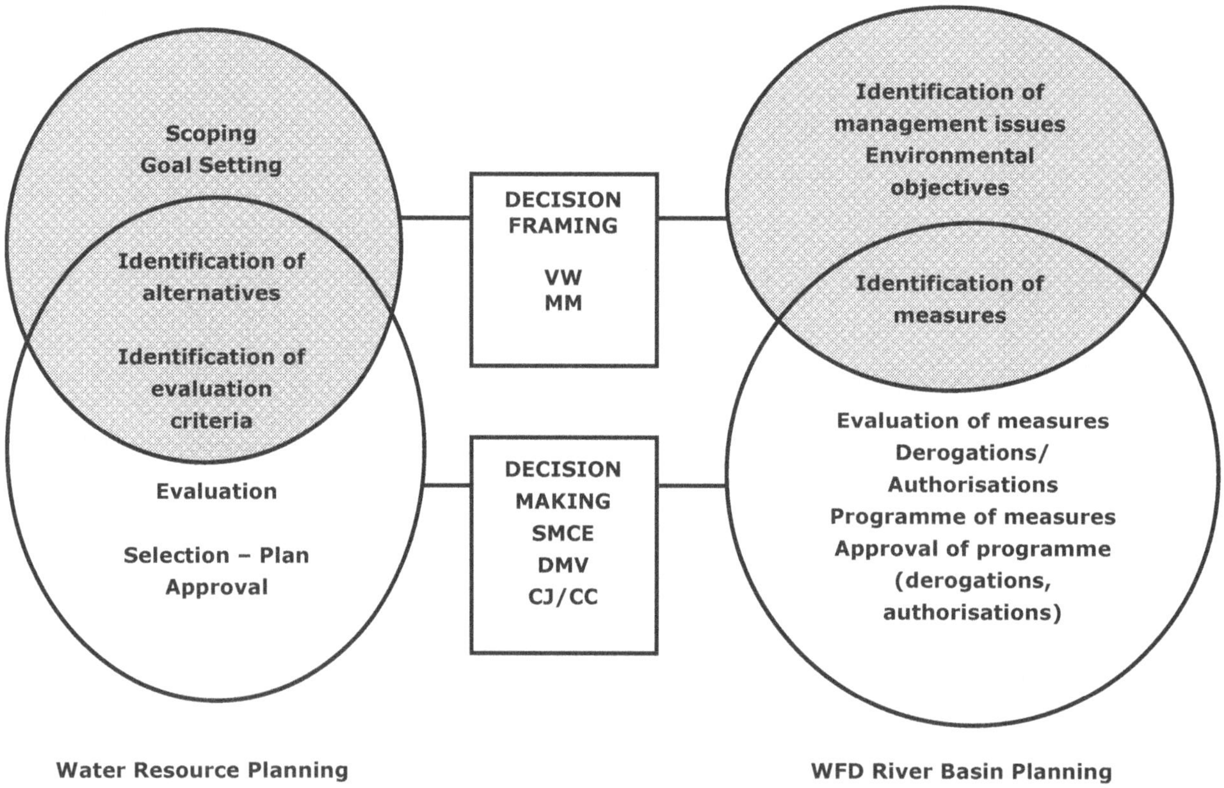

Figure 4.1 Decision Stages in water resource planning and applicability of the Toolkit[6]

VISIONING WORKSHOPS (VW)

What is a VW?

VW is a 2 to 2½ day workshop event consisting of two activities (roughly corresponding to the first and second day, respectively): vision-making and idea generation. In vision-making participants articulate, deliberate and finally agree on a vision statement about the issue/system under question for some time in the future (e.g. water management in river basin "x" in the year 2020). In idea generation participants propose specific solutions to achieve their vision. The best ideas (proposals) are voted and then participants elaborate on implementation details, ideally ending up in an Action Plan (with a time table for activities implementation and allocation of responsibilities between stakeholders).

In the future search workshop variant of a VW, vision-making builds on a number of common activities that serve to reveal to the participants their shared "common ground" and their inter-relations. In the scenario workshop variant (which formed the basis for the design of the ADVISOR VW) deliberation over the future vision builds on four alternative scenarios pre-prepared by the organisers. The scenarios provide alternative images of the future of the system.

Assumptions of VW

The underlying assumption of a VW, partly from insights in psychology and organisational science, is that by working on finding a shared vision and by liberating discussion from the burden of the present and focusing on the future, a "widest common ground" can be found without forcing or compromising. The establishment of a desirable and shared future image is expected both to energise idea generation and to act as a beacon/guide upon which to design actions. VW views human behaviour as decisively driven by perception of long-term outcomes. For a critique of these assumptions see Shipley (1997).

[6] Note that in terms of the timing of river basin planning as set in the WFD, derogations from environmental objectives of the WFD and authorisation of projects may take place at an earlier stage together with the definition of the environmental objectives.

How to?

Figure 4.2 shows the basic stages of preparing and implementing a VW using scenarios (partly based on the European Awareness Scenario Workshop methodology). The scenarios are drafted and submitted to the participants well in advance of the meeting and a "lively" presentation (photos, video, drama, etc) reminds them in the opening of the workshop. The scenarios should be sufficiently distinct, all "desirable" but from different viewpoints/worldviews and feasible (though not "business as usual"). Distinguishing the scenarios in terms of a few key dimensions (e.g. type of technologies used, public vs. private responsibility, etc) can be helpful. Scenarios are provided to participants both in technical form and in an easier, "experiential" format (e.g. a story about the everyday life of a family after 20 years, the front-cover of a newspaper same day twenty years after).

Scenario Workshops are typically conducted with around 30 participants. However, except from facilitation experience/capability and space availability, there is no other constrain to the number of participants. Future Search Workshops have been managed successfully with 80-100 participants. Workshops are divided in Assembly and Group sessions (5-8 participants max. per group). In the afternoon before the main event, participants are introduced to goals and activities of the 2 days. "Common ground building" or "ice-breaking" activities can be used to introduce participants to the philosophy and working methods of the Workshop. In the first day, participants are divided in homogeneous groups of stakeholders sharing common interests. Deliberation may be left more open (scenarios used only as a reference point) or be structured around the pre-provided scenarios (with guidelines on how to rework them). "Stakeholder group visions" result and are presented and "defended" (slogan and statement) to the assembly. Facilitated deliberation takes place leading to synthesis and a shared vision. The second day participants are mixed in heterogeneous groups and deliberate on ideas for the implementation of specific goals stemming from the vision. Best voted ideas from each group are presented and explained by the Group to the Assembly. A voting of all ideas then takes place; the best five (or more) are shortlisted for further elaboration in groups. Participants join then voluntarily any idea group in the implementation of which they wish to be partners.

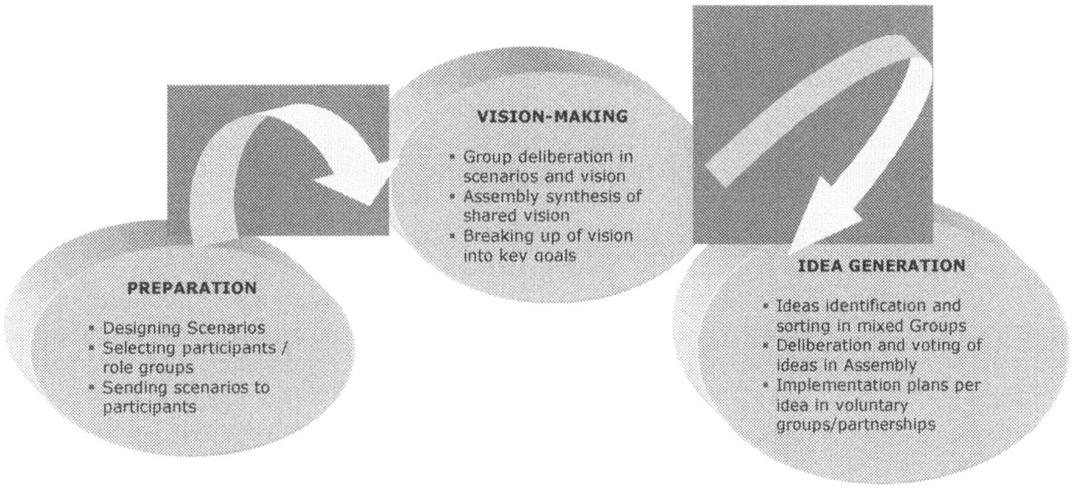

Figure 4.2 Stages of a Visioning (Scenario) Workshop

Human and Financial Resources

One or two highly experienced facilitators are needed to manage assembly workings. Group work can be self-organised or managed by the two facilitators (groups can work in separated circles in one big room). A research team is needed to collect data and draft case-relevant scenarios. Main costs include personnel expenses for the above tasks, meeting costs and participant compensations. A 2 ½ day workshop can be prepared effectively with about 18 person/months work (research and secretarial support). About 6 person/months will be needed for publication/dissemination of results and follow up activities.

Strengths and Applicability

VW are particularly good platforms for an open, initiating deliberation between stakeholders. The discussion about the future reveals clearly to participants differences in viewpoints as well as common ground. Hence, VW are particularly suited for the **scoping phase** of water resource planning. In the WFD, VW can be used to deliberate on the important management issues in the basin (e.g. a vision for the basin can be debated and then the second day devoted to see what is lacking in order to achieve it). The vision statement also can be used directly as a planning goal statement (i.e. goal setting).

The second phase is exceptionally productive in generating new ideas and can thus contribute to the identification of alternatives as part of a planning process. In a WFD context, the second day of the VW (after having defined the basin statement and goals) could be used for example to identify measures that could serve to achieve the objectives/gaps. Of course, further analysis should be undertaken to see whether the proposed measures can indeed contribute to objectives. Nonetheless, there is also a potential of innovative solutions ("information from"), neglected by "consultants", coming out from participants.

There is a potential to modify the basic VW structure to introduce a session whereby the vision statement and the goals are elaborated into more specific evaluation criteria. The VW process seems well suited for such a task as there is detailed deliberation over goals in the first day. Nonetheless, such a methodological change has not been tested in practice. Potential problems could arise in moving from the general and more easily acceptable vision statement to more concrete objectives and decision criteria.

Weaknesses and Limitations

The duration of VW is too short to allow for an in-depth deliberation of critical, conflicting issues. The structure and short duration of the process also does not allow a detailed use of scientific information in deliberation. Differences (in group, assembly) are resolved with voting or incorporated in final statements, leading to general and all-encompassing visions and loosely defined ideas. A VW cannot address the "thorny" implementation issues where conflicts occur and trade-offs are necessary. It favours the search for a common ground leaving conflicts behind and searching for win-win opportunities. Such opportunities may not always exist. For these reasons, and despite the suggestions of VW advocates which tend to obscure this limitation (partly because their experience comes from "private organisation" cases, where differences and conflicts are less pronounced), these guidelines do not recommend to use a VW when the deliberation/decision goal is to get a detailed Action Plan or to resolve specific, long-standing conflicting issues and reach a specific decision.

Although a VW may not lead to an agreed Action Plan, still it brings together stakeholders that might not have had the opportunity to communicate and exchange information in the past, thus tapping on an unrealised partnership potential that may well result in the formation of one (or more) new partnerships for projects of shared interest ("win-win" opportunities). The idea elaboration task of the second day provides a good opportunity to initiate such partnerships. Furthermore, some of the drawbacks of a VW could be overcome by extending the process and allowing more time for the deliberation to "mature". An information session could precede the VW, the idea generation and evaluation tasks of the second day could be separated and extended to two or even more days, VW could be combined and followed by other deliberative decision techniques, etc. In other words VW can be a useful component of a wider IDDP but can not deliver a decision alone.

Further reading on Visioning Workshops

Information "clearing house" about the European Awareness Scenario Workshops methodology at: www.cordis.lu/easw

Okubo D. (1997), The Community Visioning and Strategic Planning Handbook, The National Civic League, Denver, CO.

Street P, 1997, "Scenario workshops: a participatory approach to sustainable urban living?" Futures 29 (2) 139-158.

Weisbord M, Janoff S, 2000 Future search. An action guide to finding common ground in organizations and communities (Berrett-Koehler Publishers, San Francisco)

Shipley R, 1997, "Visioning in planning: is the practice based on sound theory?" Environment and Planning A 34 7-22.

Box 4.2 Visioning Workshop for the Sustainable Management of Water in the island of Naxos, Greece)

Where?

The island of Naxos is located 103 n.m. SE of Athens in the island complex of Cyclades. It has a permanent population of 18,000, increasing in July-August by 15-30,000 tourists and visitors. It is divided in two municipalities, a coastal, tourism based one ("Naxos") and the mountainous, traditional "Drimalia" living by cattlery and a local agricultural production.

Why?

- **Water conflicts** between tourism and agriculture and between Naxos and Drimalia in dry years summers;
- Tension between the two municipalities over **allocation of water** from a newly constructed reservoir;
- **Lack of long-term water resource planning**; responsive management to demands as they arise (absolute priority now to coastal tourism).

The VW was conceived as a method for **initiating public debate** over long-term planning of water resources and for **broadening the dialogue** between the competing parties and the community at large.

Proceedings and Results

The Workshop took place on the **1st and 2nd of November 2003** and was attended by **36 selected participants.** Awareness, dialogue and agreement on an Action Plan were the initial goals. In the first day, participants were divided into four homogeneous groups: policy makers and economic actors from Naxos, policy makers and economic actors from Drimalia, scientists/experts and citizens/community groups. Participants were provided in advance with four water-development scenarios for the year 2020. The "business-as-usual" scenario (**S1**) was for growth of mass tourism served by big water works. The other scenarios were: (**S2**) economic modernisation of the island, globalisation-fuelled growth with use of new water technologies and administrative approaches; (**S3**) balanced development, environmental protection, an emphasis on water conservation and small-scale, appropriate technologies; and (**S4**) radical "ecology" with self-sufficiency, communitarian self-organisation and dramatic reduction of water consumption. Scenarios were provided both in a technical format (including data tables and spatially differentiated demand forecasts) and in a "user-friendly" form of imaginary letters written by visitors to Naxos in the summer of 2020.

Each group reached a vision, which was then debated with other groups in the assembly. Helped by an experienced facilitator participants reached a shared vision that some of them described as a combination of S3 with a touch of technology from S2 and autonomy and self-sufficiency from S4. They stressed a **diversified island economy** where income and job opportunities for the young would be provided by a soft and qualitative tourism development and exploitation of competitive advantage in quality agricultural products. Water in **sufficient quantity and quality** would be secured primarily through **water conservation** and new water works, both technologically "state-of-the-art" and based on "traditional knowledge". Finally, they debated the organisational structure of the water sector favouring decentralisation, but without agreeing on the appropriate division of state, regional and local competencies.

On the second day, participants worked in four mixed thematic groups stemming from first day's vision: **water quality**; **water supply**; **water conservation**, and **institutional design**. Ideas were prioritised by voting. Each thematic group voted for the three most popular ideas. Then the assembly voted and ordered the 12 final ideas. More than 60 ideas were recorded with the three scoring highest being: school **education programmes** for water-saving, preservation and repair of rural **land terraces** to control rain-water flow, and establishment of a **laboratory** to analyse water quality. Participants proposed to distinguish in planning and allocation between water for local and water for tourist purposes.

Discussion

The workshop achieved an unprecedented **dialogue** between conflicting parties in the island. A comprehensive list of alternative proposals was put down that could potentially be used as a basis for further deliberation and decision-making. The vision statement also sets the basic goals for water resource planning in the island.

The VW fell short though of producing an Action Plan, as participants felt they lacked the necessary knowledge and **Information**. Voting of solutions was refuted by many participants as they felt the group was not representative and **lacked democratic legitimisation**. The conflicting issues of the allocation of water from the newly constructed reservoir and of allocation between tourism and agriculture in the summer month were disputed but not resolved. Discussion instead focused on alternative solutions (water conservation). Still, the allocation of available water remains an important issue and highly contested issue in the island.

MEDIATED MODELLING (MM)

What is MM?

A MM process consists of a series of **modelling workshops** where a broad stakeholder group collaborates in the development of a simulation model, which is aimed at scoping and exploring innovative solutions to complex and dynamic problems. A mediated modeller, using visually oriented modelling techniques and software, facilitates the workshops. Apart from the modelling workshops – the core platform for deliberation– MM involves **preparatory** (identifying and setting-up the stakeholder group, conducting introductory interviews and preparing a preliminary model) and **follow-up activities** (interviews or questionnaires, reports and model tutorial).

Two types of models are developed: qualitative and quantitative. **Qualitative models** consist of conceptual models depicting the main variables (and associated relationships) that describe a dynamic problem. These models may not be simulated and are only useful for problem definition and conceptualisation purposes. **Quantitative models** are achieved after the quantification of the model variables and require the synthesis of the knowledge/information disclosed by the participants into a computer-based model. These models are then used for policy analysis (typically during the final workshop) by comparing and evaluating the results of different "simulation runs", which correspond to alternative "what-if" scenarios.

Assumptions of MM

MM draws on the principles of the System Dynamics. System Dynamics is a problem-oriented method in which a given system is defined in terms of the underlying structural relationships. System Dynamics (and hence MM) assume the presence of structural cause-effect relationships underlying observed phenomena. It also assumes the ability to disentangle phenomena into a set of definable systems/variables. The tangible goal of MM is to construct a computer-based simulation model at the "scoping level", i.e. a high-generality and low-resolution model that serves to increase the understanding of the dynamic behaviour patterns at play rather than predicting a precise outcome. Eventually, the scoping models may be further iterated to more detailed and complex research or management models (van den Belt, 2004).

How to?

In a MM process the main interactive model-building stages include: problem definition, conceptualisation, specification, and policy analysis. Each consecutive stage is usually developed in distinct workshops, although the tool is flexible and variations in the process can be easily introduced. The length of each workshop depends on a number of factors (*e.g.* number of participants involved, desired model detail, problem complexity, etc) although it typically varies from half a day to two full-day meetings. The optimum number of participants for a MM process is about 20-30, although these limits are merely indicative. Past MM projects have been conducted successfully with groups of 5-10 participants (in small-group sessions), while others registered more than 40-70 participants (in plenary sessions).

Figure 4.3 presents a logical sequence for a MM process aimed at the development of a quantitative model, where the basic modelling stages are performed in four MM workshops. The model building stages should be complemented by several tasks that are performed behind the scenes of group modelling. Before the first workshop, preliminary interviews allow for collecting relevant background information from the involved actors. Results from these interviews may represent a point of departure for discussing the issues addressed by the model.

In the second workshop, model conceptualisation consists of eliciting the model's variables from the participant group in order to draw a stock-and-flow diagram that represents the structure of the problems at stake. In the following stage, participants contribute to the formulation of the model's equations so that the computer can simulate the model. To facilitate this task, behind the scenes of the third workshop, the modelling team should collect data for each variable, including the relevant documentation. The model validation procedure follows a set of tests, which may be performed solely by the modelling team. During the last workshop, alternative management scenarios are simulated and discussed with the participant group. Follow-up activities include the distribution of deliverables and the formal assessment of procedural outcomes (e.g. using questionnaires for evaluation of results).

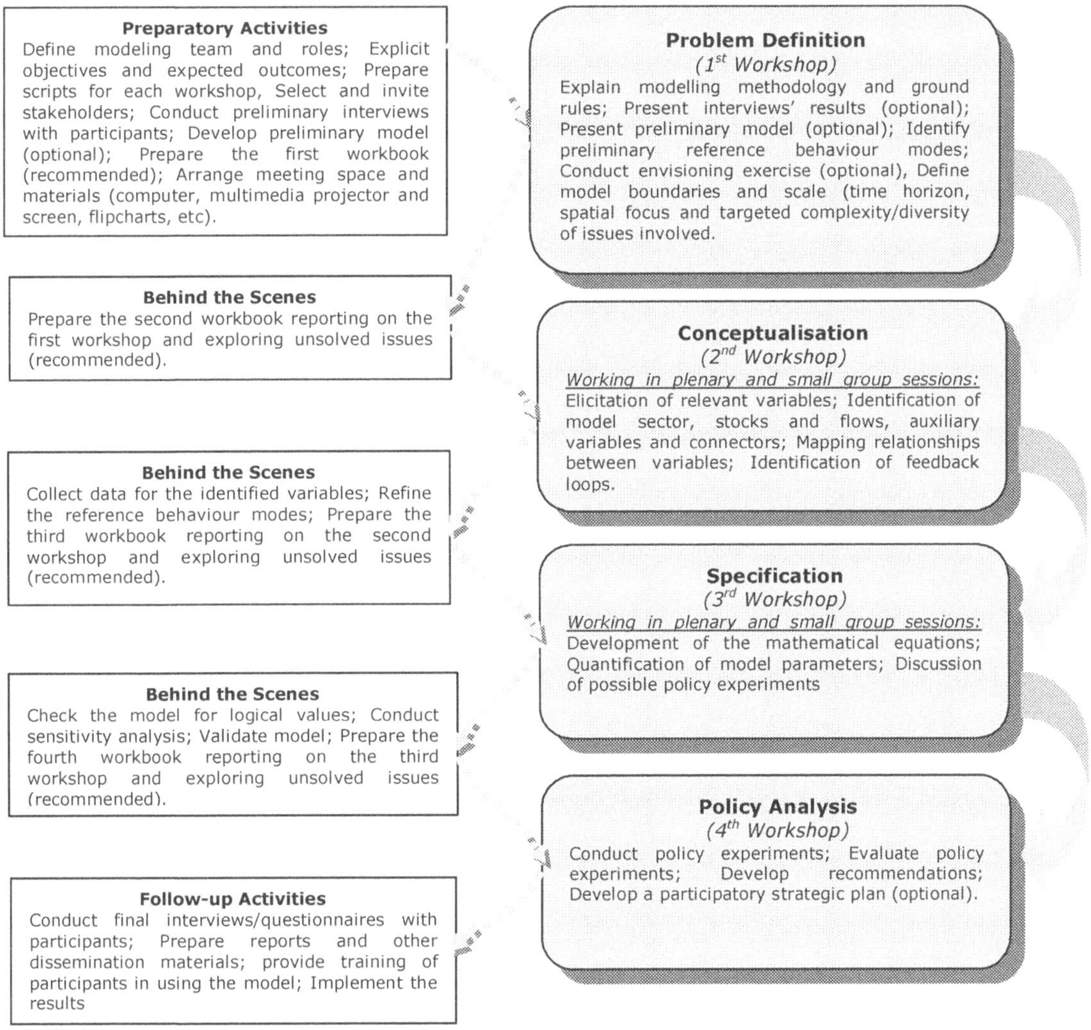

Figure 4.3 Stages of a Mediated Modelling process

Human and financial resources

Regardless of the type of partnership established, the modelling team should require at least one person within the responsible/sponsoring organization for direct contact throughout the process. This person serves as the process *gatekeeper*, i.e., someone who usually helps to initiate the process and assists in framing the management problems and identifying the appropriate participants. Other roles to be assured by one or more persons include: a *facilitator* - responsible for eliciting knowledge and conducting the group discussions; a *modeller* – who develops the model in the computer; *a process coach* - reports to the facilitator on strategies to keep the group effective; and a *recorder* - writes down and sketches the important parts of the group proceedings. Indicative human resources investment ranges from 18-24 person/months work (including preparatory, modelling and follow-up activities). Cost categories include personnel expenses for the above roles and workshop organization/distributed materials costs.

Strengths and applicability

MM allows for structured deliberation at the **scoping phase** of water resource planning. The logic and pre-defined sequence of modelling tasks shapes the backbone of the agenda for the participatory workshops and creates a shared language, shifting attention away from rigid points of view. In doing so, MM invites stakeholders to present their knowledge on the subject in an open and constructive manner, promoting team learning and knowledge integration. MM is also expected to raise the level of consensus among the group, by establishing a shared big picture first and through the task of model specification. The model building process encourages participants to focus on the leverage parameters affecting decisions. The MM outputs may be further synthesised in a strategic plan thus serving the purpose of **goal setting**. In the final workshop, during the policy evaluation stage, MM may also contribute to the planning step of **generation of alternatives**. MM's collaborative process levels the playing field during this stage by allowing all participants to propose alternatives and then focus on the analysis of their consequences over time and across different model sectors. Yet an unexplored application of MM, the formal **evaluation of alternatives** and clarification of trade-offs may be incorporated at the last stage of the MM process, contributing to reaching consent on a common decision or set of actions.

Weaknesses and limitations

MM is very flexible on some important participatory elements (e.g. length of meetings, accommodation of large groups) and deliberative applications. Nevertheless, some important limitations of MM arise and its use is discouraged when: there does not exist an experienced mediated modeller; stakeholders are not prepared to work on a voluntary basis; the problems at stake are not complex or dynamic; there are strong time and resource constraints; there exists no institutional anchor for implementing results; and when the information base is scarce.

Further Reading on Mediated Modelling

van den Belt, M. (2004) Mediated Modeling: A System Dynamics Approach to Environmental Consensus Building, (ed.), Island Press, Washington, DC.

Stave, K. (2002) Using system dynamics to improve public participation in environmental decisions. System Dynamics Review, 18 (2), 139-167.

Vennix, J.(1996) Group Model-Building: Facilitating Team Learning Using System Dynamics. Chichester: JohnWiley & Sons.

Videira, N., Antunes, P., Santos, R., Gamito, S. (2003) "Participatory Modelling in Environmental Decision-Making: The Ria Formosa Natural Park Case Study", Journal of Environmental Policy and Management, 5(3), pp.421-447.

SOCIAL MULTI-CRITERIA EVALUATION (SMCE)

What is SMCE?

Multi-criteria evaluation can be defined as a method used to **support decision-making** processes with the analysis of a number of **alternatives**, taking into account **conflicting interests** and **multiple criteria**, usually including economic, social and environmental factors. There are several possible multi-criteria evaluation platforms, as regard to the position and role of the analyst in the process and the concrete model that is used. In the ADVISOR project the **NAIADE** model was employed. NAIADE allows for two types of mutually enriching evaluations. The first is a **multi-criteria analysis** based on the score values assigned to each alternative according to the identified criteria to produce a matrix (known as the **impact matrix**). The second is a **equity evaluation**, which analyses the value judgements of the stakeholders involved in the evaluation process for each alternative using another matrix (known as the **equity matrix**) and the possible formation of **coalitions** (stakeholder groups who defend one of the proposed alternatives). The two matrices provide a platform for deliberation between the stakeholders with the potential for convergence and consensus building among one or more alternatives.

In **Participative MCE**, evaluation is the outcome of an assessment by a panel of participants. However, a participatory process can not overcome the issue of weighting the criteria (i.e. assigning the relative "weight" of each actor's preferences). Munda (2004) proposes to move from a participative to a **social multi-criteria evaluation.** The latter recognises the importance, but also the limitations of participatory approaches. **Social MCE** emphasizes the **cyclic nature** of all stages, **transparency** as an essential component of the evaluation and **reflection** and **mutual learning** between researchers and participants. To this end, it utilises a set of social science techniques to support the MCE process. These include **institutional analysis** (analysis of regulatory framework, media, economic and political processes), in-depth **interviews**, **questionnaires**, **participant observation**, **opinion polls** and **focus-groups** techniques.

Assumptions of SMCE

Multi-criteria evaluation has its epistemological foundation in incommensurability of values, averting the need for all the values to be measured on a monetary scale. Multi-criteria models make it possible to consider a large number of data, relations, criteria and objectives which are generally present in a specific real-world decision problem, so that the decision problem at hand can be studied in a multidimensional way. Multi-criteria evaluation's principal aim is not to discover a solution, but rather to construct or create something which is viewed as liable to help an actor taking part in a decision process to shape, and/or to argue, and/or to transform his preferences, or to make a decision in conformity with his goals.

A multi-criteria evaluation rests on the assumption of rationally distinguishing alternatives from objectives/criteria and on the independence of the evaluation objectives/criteria. It thus rests on a rationalistic view of human behaviour which assumes that means can be meaningfully distinguished from ends, and that different ends are separable and independent.

Box 4.3 Mediated Modelling to Support Public and Stakeholder Participation in Water Resources Planning and Management: The Baixo Guadiana Experience, Portugal

Where?

The Guadiana river basin (total drainage area of 66800 km^2) is shared between Spain (83%) and Portugal (17%). The permanent population in the Portuguese side has decreased in the last decades and was of 230,000 in the year 2000. The lower end of the Guadiana's Portuguese side is referred to as "Baixo Guadiana". It overlaps four municipalities. The MM project was hosted by the Natural Reserve of the Salt Marsh of Castro Marim and Vila Real de St.º António (NRSCMVRSA) (2089 ha designated in 1975).

Why?

- In the Baixo Guadiana, water plays a strategic role in the development of several activities such as domestic water supply to half of the Algarve region, fisheries, agriculture, salt making industry and tourism. The Baixo Guadiana also has a high nature conservation value;
- Conflicts of interests and values are present in this area; Tension between landowners (and other project promoters) and the managers of the Natural Reserve of the Salt Marsh;
- Lack of common understanding of the pressures and impacts in the Baixo Guadiana and their linkage to the achievement of the WFD objectives (including those for Protected Areas); Lack of participatory and integrated strategy for the area.

Proceedings and Results

The MM process included tasks developed along 3 modelling workshops, complemented with activities occurring "behind the scenes" of group modelling (e.g. the development of workbooks reporting on intermediate results and methodological information). The process of supplying and communicating quantitative information in a MM process is often pointed as a methodological limitation. In the Baixo Guadiana case the application of a Quality Assurance protocol was tested.

In the beginning of the process, a stakeholder analysis allowed for identifying the preliminary list of participants to invite (86 actors). Those returning positive answers were interviewed before the first workshop. This first meeting (13 February 2004) was designed to achieve three main objectives: scoping out the pressures and impacts associated with the current problems in the Baixo Guadiana, facilitating the development of an integrated qualitative model (a causal diagram) for those problems and raising awareness among the group on the challenges of participation in the implementation of the WFD. This workshop was attended by 57 participants. The group included 43 representatives from 29 organisations (national water authority, two protected areas, regional directorates for the environment, agriculture and economics, local municipalities, regional development associations, environmental associations, producers' associations, research institutes and tourism developers) and 14 property-owners. At the end of the workshop, participants reached a shared view which integrated issues associated with water salinization, increase of sediment inputs, opportunity costs of landowners due to nature conservation restrictions, and development of sustainable forms of agriculture and tourism. In the second workshop (29 March 2004) participants were invited again to work in two smaller groups (working for a half day each), for initiating (conceptualization tasks) the development of a quantitative model addressing the main problems identified by the participants during the previous workshop. The third and final workshop (2 June 2004) gathered, for half a day, a total of 20 participants. Using the model defined in the previous workshop (which was quantified by the modelling team in-between these workshops, according to the information supplied by participants) it was possible to simulate and analyse alternative management scenarios for the area. For example, the model allowed participants to discuss the effects of implementing different projects which have been proposed for the Baixo Guadiana (e.g. restoration of the saltmarsh, construction of a wastewater treatment plant, recovery of salt marshes, etc). State indicators such as a "natural capital index" and "water quality" allowed for comparing these effects. Finally, the process culminated in the definition of a list of strategic objectives and measures foreseen by the group to support integrated and participatory planning and management of the Baixo Guadiana. Participants were also asked to return a questionnaire to evaluate the results of the MM process.

Discussion

The qualitative modelling approach used in the first workshop (where more participants were present) seems to be well suited for early problem scoping stages of a planning/evaluation process. This approach worked very well in the development of an integrated view of the problems, pressures and impacts in the area, allowing to structure and synthesizing group discussions with the use of simple cause-effect relationships. As the participation rate decreased in the subsequent stages of the experiment, the quantitative model fell somewhat short of including all the relevant viewpoints previously identified. Nevertheless, only the quantitative model equipped participants in the final workshop with a tool for analysing the consequences of different scenarios. This model successfully supported the strategic assessment of alternatives and facilitated the analysis of trade-offs.

Perhaps even more important than the developed models and the preliminary drafting of strategic objectives and measures, the MM process allowed for integrating knowledge and promoting **team learning** among the elements of a group that seldom gather for deliberation. Additional information on the case study is available from http://ecoman.dcea.fct.unl.pt/projects/advisor/iddp/.

How to?

First, **institutional analysis** and **social research** are applied to determine the **scope of study** and to **identify the stakeholders and interests** involved (Figure 4.4). Issues examined include historical and geographical conditions, normative context, economic and political processes, through analysis of documentation and national/local media, surveys, interviews and participant observation.

These analyses are also necessary in order to frame the problem at hand and to start identifying the alternatives of action and criteria evaluation. These tasks can be assisted by conducting **in-depth interviews** and **written questionnaires**. The interviews may be conducted according to a set questionnaire, which however has to be flexible enough to allow stakeholders to freely express their opinions. The interviews could take around an hour although this can be adapted to the time availability of the stakeholders. It is recommended to record the interviews to allow a detailed analysis of their content. The information provided by the stakeholders in the interviews can be elaborated on in a written questionnaire to be filled out by them after the interview to refine some questions which may need clarification.

On the basis of the alternatives and criteria identified in the previous stage, the NAIADE model is applied to evaluate the alternatives, firstly, according to the different criteria –**impact matrix**-, based on data from specialized literature and technical reports. Concerning some social/institutional criteria (e.g. public acceptance of the alternatives) it would be useful to conduct an **opinion poll** among the inhabitants of the area. Secondly, NAIADE is applied to evaluate the alternatives according to the positions of the stakeholders **-equity matrix-** as well as the potential formation of coalitions among the actors to defend or veto a given alternative (dendrogram of coalition's representation).

Collective and open discussion and deliberation over results then takes place in a **focus group**, with a view to obtaining feedback, sharing and clarifying the information collected and discussing the results achieved up to that point.

Human and Financial Resources

Human resource requirements for the various analytical tasks (institutional analysis, interviews and surveys, management of focus groups, reporting, etc) are approximately 20 person-months of experienced researchers with multi-disciplinary background. Other expenses include standard organisational costs (travel, consumables, and secretarial support) and costs for at least one focus group meeting. The application of the methodology requires considerable technical information (whether quantitative or qualitative) to evaluate each of the criteria. Costs of additional research, data gathering and analysis can be considerable raising significantly the cost of the exercise. This requirement can be one of the main constraints of the exercise, although the flexibility of the tool to process and incorporate different types of information, including linguistic expressions, somewhat lessens this problem.

Strengths and Applicability

Like VW and MM, deliberation between stakeholders over alternative solutions has the potential of revealing new innovative ideas, previously not considered. Furthermore, the cyclic nature of SMCE is flexible and allows the addition of new alternatives as the learning process proceeds (c.f. VW where identification of ideas is a one-off exercise).

SMCE is designed explicitly and hence is particularly well suited for resolving conflicts and finding convergence points between stakeholders' viewpoints. It treats multiple values in the evaluation explicitly, without attempt to "homogenise" them (incommensurability). SMCE has also the capacity to manage quantitative and qualitative information for the evaluation (c.f. for example VW). Hence, it is a tool particularly well suited for decision aid in the evaluation phase of planning (see Box 4.4). In the WFD river basin planning process it can be used for the evaluation of the alternative measures (or combinations of) and complement hence a cost-effectiveness analysis. Potentially, it can also be used for authorisations/derogations. Two or more options (e.g. a "new modification" vs. an alternative such as demand management) can be compared upon the range of exception criteria defined in the WFD text (e.g. costs, beneficial objectives, technical feasibility, environmental impacts, etc). Note that in such an application, criteria will be pre-defined (c.f. SMCE where participants define alternatives).

Weaknesses and Limitations

SMCE presupposes that each stakeholder has the same degree of power in the decision, making the outcome less realistic. Main limitations of a social or participatory MCE stem from the stakeholder-based participatory structure (see Chapter 3) and can only be overcome by contextualising the process in broader legitimate decision structures and combining with other participatory processes.

Another drawback of SMCE (relative to VW or MM) is that the construction of the criteria-based evaluation matrix is highly **complicated** and depends a lot on the scientific support (as do institutional and other sociological analysis).

Adding to this the often considerable limitations in terms of data and the result may be a lack of confidence and acceptance by participants of the results. The better and more comprehensively designed the process the more the acceptance by participants will be. It is important to clarify that SMCE is not supposed to provide *the* decision, but to **aid and structure** the debate that leads to the decision.

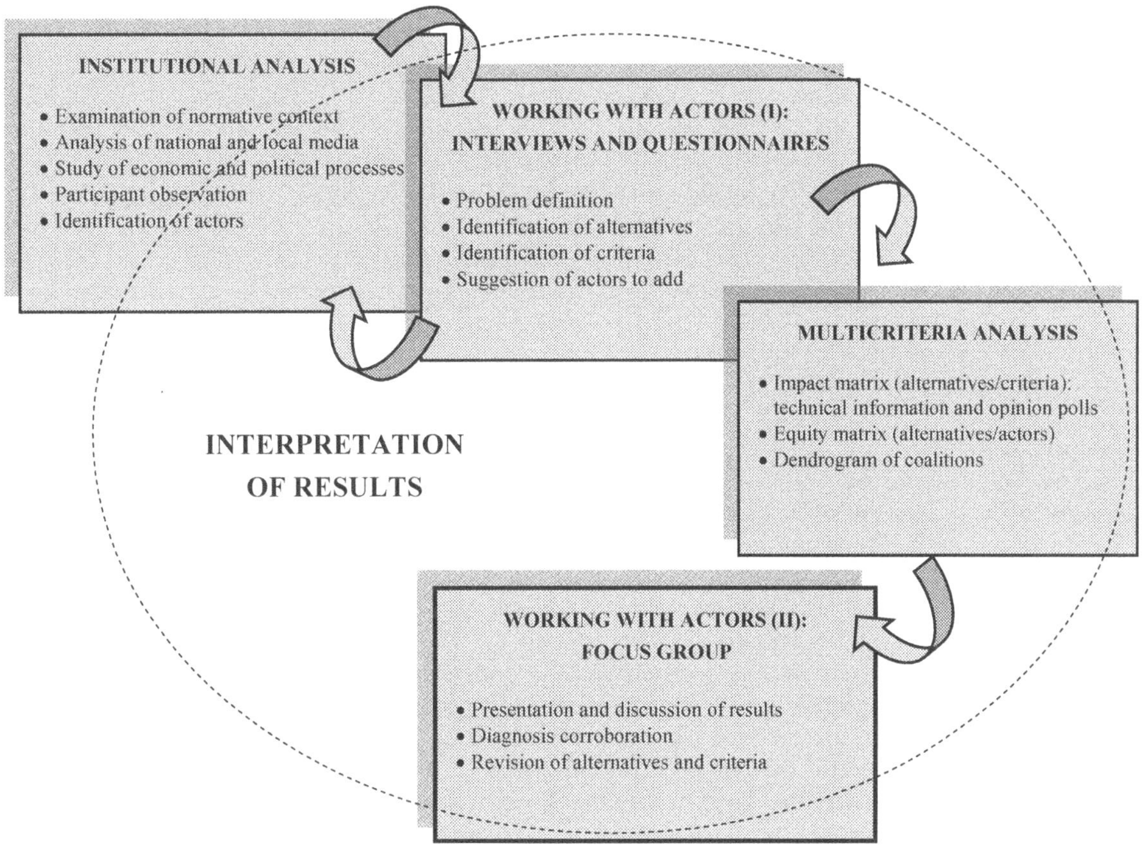

Figure 4.4 Stages of a Social Multi-Criteria Evaluation

Further reading on Social Multi-criteria Evaluation

Corral Quintana, S. (2000) Una metodología integrada de exploración y comprensión de los procesos de elaboración de políticas públicas, Universidad de La Laguna, España. ISBN 84-688-4314-8.

De Marchi et al. (2000): "Combining Participative and Institutional Approaches with Multi-criteria Evaluation. An Empirical Study for Water Issues in Troina, Sicily", Ecological Economics, 34: 267-282.

Guimarães Pereira, Â. Blasques, J., Corral Quintana, S., Funtowicz, S. (2003), Tools To Inform Debates, Dialogues & Deliberations: The GOUVERNe Project at the JRC. European Commission: Joint Research Centre, Ispra. EUR 21189 EN

JRC (1996): NAIADE Manual & Tutorial. Ispra: Joint Research Centre.

Munda, G. (2004): "Social multi-criteria evaluation: methodological foundations and operational consequences", European Journal of Operational Research, vol.158/3, pp. 662-677.

Munda, G. (1995): Multicriteria Evaluation in a Fuzzy Environment. Berlin, Physica-Verlag.

Box 4.4 Social Multi-criteria Evaluation of water management alternatives for improving water supply in Costa del Sol Occidental, Malaga, Spain

Where?

The Costa del Sol is located in the Region of Andalusia in the southern Mediterranean Coast of Spain. Population growth (341,401 inhabitants in 2001) has accelerated in recent years. Population triples due to tourists in the summer months and doubles during the rest of the year. The area of the study comprises of eleven municipalities and roughly coincides to a catchment.

Why?

- Spell of **droughts** that have highlighted **competition and conflict** between different water uses.
- **Uncontrolled** increase in certain water uses (especially tourism related).
- **Lack of anticipatory planning** by the competent authorities aggravating problems and conflicts.
- **Administrative rivalries** between central government authorities and the regional authorities over water management powers and failure of the River Basin Authority to establish adequate channels for public participation.

SMCE was conceived as a useful tool to explore alternative options over water supply, account for the diversity of judgement values and interests at stake regarding water and contribute to the resolution of users and institutional conflicts.

Proceedings and Results

First, the research team carried out an analysis of the normative context, media, economic and political process, as well as 16 in-depth interviews and questionnaires in order to identify the stakeholders and to define the problem at hand. Researchers also participated in two important open debates held in Malaga for a better observation of the context. The diagnosis concluded that the majority of the stakeholders regarded the Costa del Sol Occidental as an area beleaguered by insufficient water management (especially concerning groundwater and effluents), lack of forward planning, inadequate coordination among the authorities responsible for water management and those responsible for sectoral and spatial planning as well as weak information and participation mechanisms.

Interviews and questionnaires were also used to identify alternatives – which included both actions aimed at increasing available resources and measures to improve demand management - and criteria evaluation. Most of the stakeholders had more trouble identifying criteria to evaluate these alternatives, beyond the usual categorisation into economic, environmental and social criteria. This made the research team realise the need to provide a more studied proposal of criteria as a starting point. The alternatives and criteria considered resulting included:

Alternatives: *Heightening of the existing water supply dam; use of desalinated water; reuse of waste water; modernisation of irrigation systems; rationalised use of groundwater; improved efficiency and water savings in the urban water supply; territorial policies to control urban development; and non-intervention.*

Criteria: Implementation costs; operating costs; effect on employment; effect on economic activity; impact on the ecological status of water systems; impact on other ecosystems; visual impact on the landscape; degree of institutional difficulty; degree of social acceptance; equitable distribution of costs and benefits; and time required to fulfil the established objective.

All the information gathered in the previous process were used to implement the following stages using the NAIADE model: construction of impact matrix (based on technical information and an opinion poll conducted to evaluate the 'degree of social acceptance' criteria), equity matrix (based on stakeholders' positions and information showed in the interviews and questionnaires) and dendrogram of stakeholders' coalitions. The result showed that the **demand-management alternatives** were the highest scoring and the bottom one was the heightening the La Concepción dam, the main supply-side alternative. In the last stage of the methodological framework, a half-day focus group meeting was carried out in an hotel in Malaga. As a result two key considerations were detected. First, when the stakeholders were confronted with the final list of alternatives that had been worked on, they suggested the **inclusion of new alternatives.** One of these alternatives (reforestation of catchment area) finally got the highest support. Second, in this phase of the deliberative process, there was withdrawal and **absence of powerful** actors.

Discussion

The Costa del Sol SMCE was a successful mutual learning experience for the participating stakeholders and the research team. New knowledge was co-produced. The process also contributed to clarifying the nature of conflicts and creating the conditions necessary to construct a dialogue process among the stakeholders more permanently involved and to find a meeting point between opposing positions. It also promoted active participation in all evaluation phases and enabled stakeholders to acquire knowledge of each other and their respective positions and viewpoints. The decision impact of SMCE cannot be assessed as the Costa del Sol was run on an experimental basis. There are concerns that a real life application would be constrained by inherent barriers to all deliberative processes (e.g. abstaining of powerful actors, lack of embedment in real decisions, etc). Complexity of the tool, lack of the data necessary to perform evaluations, and problems with stakeholder representation are some of the other drawbacks of the tool that were revealed in the Costa del Sol experiment, and that should be addressed for future applications. Additional information on the case study is available from http://ecoman.dcea.fct.unl.pt/projects/advisor/iddp/.

DELIBERATIVE MONETARY VALUATION (DMV)

What is Monetary Valuation?

Valuation of the monetary costs and benefits of alternative options is central in any type of water management decision. **Cost Benefit Analysis (CBA)** and **Cost Effectiveness Analysis (CEA)** are standard techniques that can be used for evaluation in the context of the WFD. Relevant guidance is provided by the CIS Working Group on Economic and the Environment. A novel combination of CBA with CEA in the formulation of a flood protection plan was also developed and tested as part of the ADVISOR project.[7] CBA and CEA can be run parallel or in combination, informing and being informed by an IDDP. Deliberative forums for example can be used to set the scope of CBA (boundaries, alternatives considered, etc), define the objectives of CEA, control the results of a CBA, etc.

The inclusion of environmental effects in a CBA requires the **economic (monetary) valuation** of environmental goods and services. Since no markets for these goods and services exist, proxy valuation techniques have been developed (Table 4.1).[8]

Table 4.1 Techniques for the Estimation of Monetary Values

Technique	Method of Value Estimation
Market Price	From ecosystem services or products that are bought or sold in commercial markets via changes in demand for products/services
Replacement Cost	From prices individual groups or societies pay to replace lost ecosystem services
Relocation Cost	From prices individual groups or societies pay to relocate assets, infrastructures or housing displaced by physical implementation of policy or threat of environmental problems
Production Function	By modelling change(s) in quality and quantity of output due to changes in factor inputs
Recreational Demand (e.g. Travel Cost)	Uses loss of tourist revenues as a proxy for the cost to society of the loss of natural features
Hedonic Pricing	From asset price changes (e.g. property value, wage differential, etc)
Contingent Valuation	By asking people to directly state a bid for specific environmental services, based on a hypothetical scenario
Contingent Choice	By asking people to make trade-offs among sets of scenarios
Deliberative Monetary Valuation	Combining contingent valuation questions with deliberation
Benefit Transfer	Transfers existing value from studies completed for another location

Drawbacks of CBA, CEA and monetary valuation techniques include:

- Considering only one type of value (monetary);
- Assuming constant and non-changeable preferences held by individuals;
- Taking the scope of the evaluation as given;
- Not addressing equity issues.

These drawbacks can partly be addressed by introducing deliberative aspects in monetary and cost/benefit valuation or by embedding economic valuation techniques in a broader IDDP framework.

[7] Bruisma, F. and R. Vreeker (1994), "Combined Cost Benefit Analysis and Cost Effectiveness Analysis Framework for Flood Protection: Case Study of the Rhine and Meuse River Basins" (available at Advisor website: http://ecoman.dcea.fct.unl.pt/projects/advisor).

[8] For a review see Spash, C. and K.Urama (2003), "Monetary Valuation Methods Reviewed for Relevance to River Basin Management", as part of the Advisor project (available at Advisor website: http://ecoman.dcea.fct.unl.pt/projects/advisor).

What is DMV?[9]

DMV refers to a variety of approaches through which deliberative aspects can be introduced in methods for the estimation of environmental monetary values. Two basic approaches for combining deliberation and monetary valuation can be identified.

> **Assumptions of DMV**
> DMV endorses the basic premises of the economic model, i.e. individualistic, rational, utility-maximising human behaviour. It differs however in some important aspects. Imperfect information of individuals is recognised; hence a process of reflection and deliberation can help in revealing real preferences and values. Furthermore, individual preferences are not considered as given and constant, but as both given and constructed and hence malleable through deliberation and information. DMV is essentially instrumentalist; the process of deliberation is seen as a tool for an outcome (getting an economic value) and not as important in its own right.

The first considers deliberative processes as a **complement** that can benefit, or used in combination, with monetary valuation techniques. Monetary valuation may be either **followed or preceded** by some element of deliberation. A variety of possibilities exist. Group deliberation may be used to **test the MV survey** design (e.g. wording and comprehension of questions), validate the information content and help identify design biases. A deliberative process may be used to **determine the options** (alternatives) or institutional context, upon which the valuation survey will be based. Stakeholder deliberation may also be employed to **validate the outcomes** of an environmental valuation. Deliberative processes and environmental valuation therefore may be sequential as part of a broader IDDP. For example, a citizens′ focus group may be compiled to deliberate on the design of a survey for the monetary value of restoring the ecological status of a damaged river. Analysts then will perform the survey and bring back the results for consideration and deliberation to the citizens′ group before using them for a CBA. Recommendations from the CBA should then be used in conjunction with deliberative processes recommendations in taking the decision whether the investment in restoring the river is "feasible" or "disproportionately expensive".

The second approach refers to the use of **formal deliberation concerning an environmental impact in order to express value** in monetary terms for policy purposes or for an input to a CBA. Group deliberation thus provides the instrument per se for the valuation (instead of say, a survey). For example, consider the case of a river basin authority that has to decide on the authorisation or not of a dam that will impact on the habitat of a number of rare or threatened species. A group of citizens would be selected and meet (e.g. through a Citizens Jury; see next) to discuss information about the environmental damages associated with the development. Known costs and benefits would be presented, while those pertaining to environmental damages would be deliberated. The result would then be incorporated into a net present value calculation to determine the viability (costs and benefits) of the project (and hence inform on whether an authorisation should be given).

Human and Financial Resources

Costs should more or less correspond to the sum of a deliberative process (see other tools presented in this section) plus the costs of a monetary valuation exercise (e.g. collection of scientific information, conduct of surveys, etc). Evidently, a team of economists with expertise in environmental monetary valuation is required in addition to the standard requirements of a deliberative process. A DMV therefore will have a significantly higher cost than a deliberative or a MV process alone. An advantage in practice however may be that the cost of the process can be included in the overall budget of a CBA, for which generally there are considerable financial resources (c.f. resources for participatory processes).

Strengths and Applicability

Monetary valuations and CBA are central in water resource planning and evaluation. Policy-makers will always recourse to some type of money-related information in taking decisions. The reference of the WFD to "disproportionate costs" or "economically significant species" serves as testament. Monetary valuations in the WFD will be central in the estimation of (disproportionate) costs for derogations and heavily modified waters and for the cost effectiveness

[9] Information in this section benefits from Niemeyer and Spash (2001).

analysis of alternative measures (see also CIS WFD Guidance on "Water and the Environment"). Combining MV with deliberative processes as part of a longer IDDP (1st variant of DMV) will improve the quality of decision aid/making, providing a plurality of information on the different options. Deliberative valuations (2nd variant) have the advantage that in comparison to conventional monetary valuations they allow for a certain degree of deliberation (and hence public participation) before deriving the values. Hence they can also account for broader concerns of equity, justice, uncertainty, etc.

Weaknesses and Limitations

Combining deliberative processes with economic valuation may prove problematic in so far as the feedback from the public may diverge from the economic assumptions upon which MV is based (for example, the expressed desire for community compensation unrelated to the individual). The "logic" of an MV may prove difficult to be articulated to and accepted by the public or the stakeholders, limiting thus the usefulness of their contribution. Particularly, in the case of the second variant, there are gross problems in "straight jacketing" the inherent pluralism of a deliberative process into the instrumental goal of deriving an economic value. Furthermore, in comparison to surveys which can target a representative part of the population, panels face limitations; there are concerns therefore whether the resulting monetary values are indeed representative "market proxies". It is recommended to use primarily deliberative and MV processes in parallel, one feeding into the other (1st variant). For example, outputs from MV may feed in a SMCE or a CJ, or vice versa, identification of alternatives in an SMCE or a CJ used for designing an MV survey. A deliberative valuation (2nd variant) is recommended to be used only for the allocation of public funds after the specific options have been yielded and not for the actual monetary evaluation of environmental services. For example, a deliberative forum may be run by a river basin authority to discuss on the allocation of its budget upon alternative investments or programmes.

Further reading on Deliberative Monetary Valuation

Blamey, R., James, R and S. Niemeyer (2001), Citizens Juries for Environmental Management: An Alternative to CBA? Final Report, Research Project ANU11 of the Social and Institutional Research Program of Land & Water Australia (downloadable at: www.lwa.gov.au)

Niemeyer, S. and C. L. Spash (2001). "Environmental Valuation Analysis, Public Deliberation and their Pragmatic Syntheses: A Critical Appraisal" Environment & Planning C: Government & Policy 19(4): 567-586.

TOOLS FOR COMMUNITY OPINION

The tools developed in the ADVISOR project focussed mainly on techniques where deliberation was done between groups of stakeholders (inner core of figure 3.2). The ADVISOR tools address planning issues, but are less well positioned (perhaps with the exception of SMCE under certain modifications) to address **controversial issues** (such as the authorisation or not to a new dam). A comprehensive IDDP toolkit should be complemented with tools and processes that address such situations. Citizens' Juries (CJ) (and to an extent, Consensus Conferences (CCs)) can be employed in such contexts.

What are CJ and CCs?

Assumptions of CJ/CCs

CJ/CCs reflect models of direct democracy (e.g. appointment of authorities by draw). They are based on the rationale that given adequate information and opportunity, all people, even if not professionally trained, can be trusted to take decisions regarded as legitimate and fair on behalf of the community.

Figure 4.5 shows the basic stages of a Jury process. The Steering Committee comprises stakeholders knowledgeable of the issue under discussion. They frame the goal and set of questions that should be addressed by the CJ ("charge"). Organisers undertake then the task of selecting individuals from the community in a **random yet representative way** (in terms of demography, incomes, social characteristics, etc). The Jury usually, consisting of 10-20 individuals, invites then a number of "witnesses", experts or stakeholders knowledgeable of the issues and engages in a dialogue over several days of professionally moderated hearing ensuring that all questions are answered. The jurors deliberate and at the end of the process they issue a report with key findings in response to the charge and their **recommendations**. The document is written and expressed by the jurors themselves although secretarial support may be provided.

The **CCs** process is very similar to a CJ. CCs however have been primarily used to provide community input to controversial scientific questions or debating new technologies, rather than verdicts to questions of policy. Hence the panel of citizens (10-20 people) is not representative but selected after an **open invitation** for all those interested. The citizens' panel is briefed in two weekend sessions on the issue under consideration. The panel identifies the questions to be asked to experts/witnesses. Written responses are provided and in addition a Conference, open to the public, is held lasting typically 3-4 days. Members of the public other than in the panel can also pose questions to experts. The panel then retires and issues a report with a judgement on the issue and presents it to the audience and to the mass media in a special press conference. Whereas in citizens' juries there is more room for different opinions in the final verdict, in a CCs the emphasis is more on reaching consent in the final decision. Applications of CCs have differed from those of CJ; emphasis has been primarily on informing scientists on public opinion in controversial issues in order to improve research and application practices. If random selection however is used, CCs can be used like CJ, in the discussion of controversial water management issues.

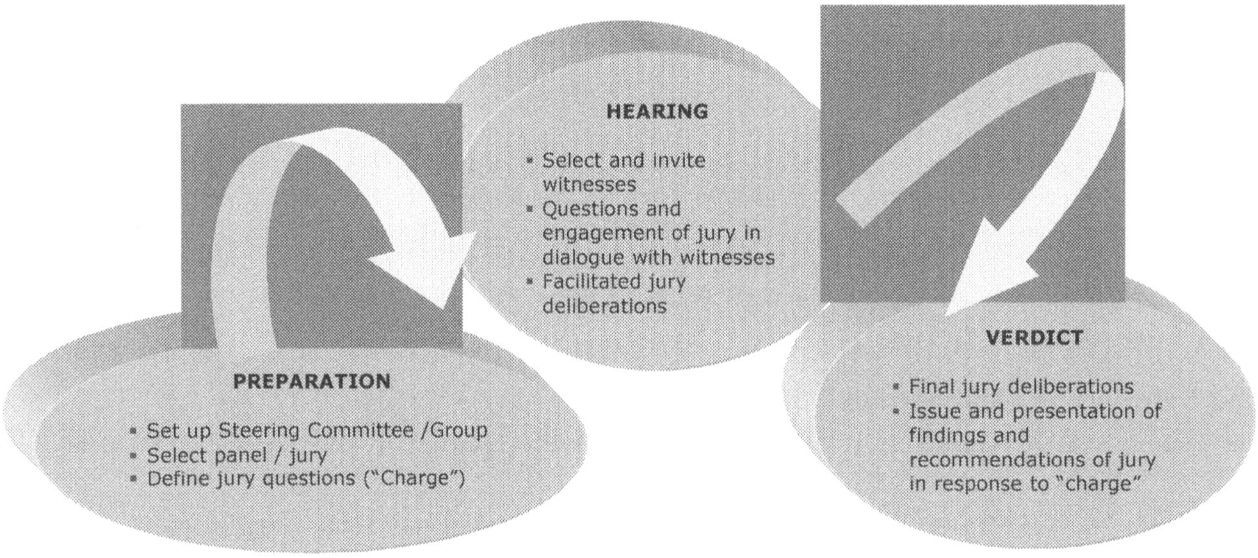

Figure 4.5 Stages of a Citizens Jury process

Human and Financial Resources

Costs include personnel for the overall organisation and support, payment and expenses of panel and witnesses (note that witnesses may be experts invited from far and need compensation as well as travel and subsistence), hire of venue, costs of facilitation, recording and transcribing proceedings and costs for any related research/analysis and publication/dissemination work. CJs have typically cost between 20 to 30.000 Euros to set up. Costs may increase depending on the level of commitment.

Strengths and Applicability

The main strengths of CJ/CCs is that they provide a direct and "fair" way for getting a community opinion and recommendations on a controversial issue. As such they are suited for providing **opinions** on the approval of water resource / river basin plans, the authorisation of controversial projects and derogations or assignment of heavily modified waters. They can aid the final stage in a longer IDDP, proceedings of the Jury/Panel informed by the results of prior research, analyses or even other deliberative processes (e.g. MV, SMCE). Alternatively a CJ/CCs can be held earlier on the planning/decision-making process. The "charge" of the panel can be to recommend alternatives for a water resource plan or project (see Brenton catchment example in Box 4.5). More detailed study and evaluation of the alternatives can then follow (e.g. through further scientific research, DMV, SMCE).

Weaknesses and Limitations

Limitations of CJ/CC relate to the inherent drawbacks of direct democratic models and random representation. Results are highly dependable on the ability of panel members (jurors) to make critical and responsible use of the information.

This might be difficult in very specific questions (e.g. derogation from ecological objectives of the WFD). **Framing** also is a much more decisive task (and hence possible source of bias) than in a VW or an MM, as the "charge" determines the content and "breadth" of the panel deliberations.

Having a "representative" panel is no easy task: identification of the population to be sampled and selection of the **representative criteria** are contested issues. In controversial issues **selection of witnesses** will be also decisive for deliberations and Panel's verdict. Note that the relatively short duration of the Jury does not allow verification of data provided by witnesses; there is a danger of decisions based on false claims (CCs partly address such problems by allowing for written responses, which are studied by the Panel before the conference with the option for new questions/clarifications). Another drawback is that witnesses who have the ability to impress may be advantaged in the process over "low-key" witnesses or witnesses with more scientifically complex arguments.

It is questionable whether a CJ/CC on a very specific issue which is not of "direct" interest to the community can attract the necessary level of interest and commitment from panel members. In certain locales many of the technical issues of the WFD (e.g. derogations, heavily modified waters, etc) may face such apathy.

A CJ or a CC has also high organisational and time/money requirements, especially with regard to the management and presentation of scientific information to the panel. Quality and effectiveness of the process depends a lot on the quality and commitment of organisers.

Therefore whereas CJ/CCs are useful components in an IDDP toolkit, and can yield a broader community opinion (and recommendations) on an issue, this does not mean that they should become *the* decisive step in the deliberation over a final plan, decision, etc (i.e. act as real "Juries"). They should complement and be complemented by other decision (deliberative and others) processes.

Box 4.5 provides two illustrations of the type of contribution that a CJ can have on water resource planning and evaluation. Note that in comparison to the examples for VW, MM and SMCE (Boxes 4.2-4.4), which included case studies carried out and tested in the ADVISOR project, the CJ examples are taken from the literature, and are not a product of the ADVISOR project. In this respect, there is no critical discussion of the experience from the application of the tool.

Further reading on Citizens Juries and Consensus Conferences

Australian research project on CJ for natural resource management: www.lwa.gov.au/downloads/PF010167.pdf

Jefferson Centre (trade marking CJ in the U.S.): www.jefferson-center.org/citizens_jury.htm

Aldred, J and M. Jacobs (2000), Citizens and wetlands: evaluating the Ely citizens' jury, Ecological Economics, 34 (2), 217 – 232.

Coote, A. and J. Lenaghan (1997), Citizens' Juries: theory into practice, London: Institute for Public Policy Research.

Crosby, N. (1995), Citizens Juries: one solution for difficult environmental questions, In Renn, O et al (eds), Fairness and Competence in Citizen Participation. Dordrecht: Kluwer Academic Publishers

Joss, S. (1998), Danish Consensus Conferences as a model of participatory technology assessment, Science and Public Policy, 25 (1), 2-22.

Box 4.5 Two illustrations of the use of Citizens' Juries on Water issues

The Brenton Catchment in Australia (www.elton.com.au/talkBronte/bronte_recs.htm)
The goal of the Brenton Catchment Project (BCP) in Syndey, Australia is to improve the quality of water in the Brenton beach by source management of stormwater quality in the cathcment. The experience is relevant to EU communities aiming to achieve bathing water quality standards in the context of catchment management as part of the WFD.

Community involvement and deliberation were "built in" from the beginning in the process of developing the BCP. A Citizens' Jury for BCP was preceded and supported by a Citizens Tele-poll which involved a directly-addressed mail-out to every household listed in the telephone book for the Bronte Catchmen). Residents were sent an information package and asked to read through it, before receiving a phone call during which they were asked to respond to a series of questions. 358 out of 877 households participated in the tele-poll (an extremely high response rate of 41%). Their responses were analysed and presented to the Citizens Jury two weeks later.

The Citizens Jury (hearing in October 2001) involved 15 citizens, recruited through a range of strategies including the Citizens tele-poll, newspaper advertisements, information stalls at community events, presentations to community groups and precinct committee meetings, as well as posters displayed across a range of community locations. Jury candidates were interviewed and asked a series of questions, to select a demographic cross-section of people with a range of environmental views, values and involvement in the local community. Jurors received detailed briefing papers and met for an informal discussion about the Jury process and a series of catchment-based activities to focus on the issues. Jurors received additional information about the issues in a series of pre-prepared reports, heard a series of expert witness presentations, questioned witnesses, and then retired to deliberate together over a series of recommendations to a range of stakeholder groups. The process was supported by two facilitators and the Bronte Catchment Project Team. The recommendations were grouped by the Jurors into the following categories: research and monitoring; community education and participation; urban planning and design; capital works and innovative projects; regulation and enforcement. Recommendations included specific proposals such as "introducing compulsory rainwater tanks systems" or "financing important low cost projects such as a public car washing space". The recommendations were presented to representatives from the local Council, and the National Environmental Protection Agency in a formal ceremony at the close of the Jury. The proposal of the Jury was the Council and the Brenton Catchment Project take up, feedback and provide a regular reporting on progress in the implementation of the recommendations. The success of the Bronte CJ as a community educative instrument process has been documented but its real decision impact on the BCP remains to be evaluated.

The Ely Wetland in the U.K. (Aldred and Jacobs, 2000; http://alba.jrc.it/valse/report.htm)
The Ely CJ was organized in July 1997 in the context of the EU-funded VALSE research project. The Jury was organized by Lancaster University with the support of an Advisory Group consisting of local councils, organizations, and residents. The Jury addressed the creation of a wetland area in the region of Fens. The experience can be relevant to local communities or river basin authorities in the EU debating restoration projects in the context of the WFD.

The Jury's charge was "what priority, of any, should be given to the creation of wetlands in Fends?" 16 jurors, lay persons with no previous expertise, representing a cross-section of the population in the Fens and adjacent areas were recruited by a market research company. The Jury sat for four days, hearing evidence from a variety of expert and other witnesses, and discussed the issues surrounding wetland creation in small groups and in plenary sessions. The discussion centered on four options, reach setting out different ways of creating wetland for different purposes, including the option that nothing should be done at all. The Jury was assisted by an independent facilitator.

The Jury's conclusions and recommendations as a whole were unanimous, although some jurors abstained or registered dissenting votes on some individual conclusions. The Jury supported the proposal of a large natural reserve (as large as 12,000 Acres) and rejected the proposal for a multi-use recreational centre. It recommended specific elements that should be included in the natural reserve (such as an agricultural/ecological research station and camping facilities) and set out also a broader policy of incentives and planning measures for the incremental development of wetlands throughout the Fens. Furthermore, as part of deliberation in the Jury over the type of agriculture in the area, a new idea of a "Fens Covent Garden" emerged and was included in the recommendations, i.e. a local wholesale centre to distribute fruit and vegetables produced in the area.

Chapter 5

The Way Ahead

This document provided some first directions for implementing IDDPs in water resource planning and evaluation. Attention was given to emphasise that an IDDP is no "magic solution" to water resource decision-making problems. Indeed, deliberative processes are ridden with difficulties. The perspective of the ADVISOR project is that these should not be seen as insurmountable limitations or as evidence against the usefulness of deliberation. Instead, these difficulties provide challenges for improving both the theory and the practice of deliberation. The main challenges for the future are listed below.

Challenge 1 – Improving / combining deliberative tools

The tools tested in the ADVISOR project are flexible and amenable to several modifications. For example, the duration of a VW can be increased, or sessions upon which participants deliberate on evaluation criteria can be added to a VW or a MM.

There exist also several other deliberative tools that can be modified and innumerable possibilities of combinations and hybrid tools. Some examples: a VW could make use of Mediated Modelling, or vice versa, a MM start with a VW. Visual models (e.g. Geographical Information Systems, etc) can be utilised in VW or MM. VW and MM could be enriched with the use of explicit conflict resolution or consensus building techniques and sessions. Multi-criteria evaluation could be applied upon deliberatively prepared scenarios. The interfaces between Citizens' Juries and Monetary Valuations could be further explored.

All these possible modifications and combinations suggest for a very promising and challenging **research programme**. This programme is not constrained to water resource planning only, but is generally important for environmental management and community planning.

Challenge 2 - Testing IDDPs in practice

The IDDP provided in this document was a conceptual synthesis from the experience from the application of separate tools in the ADVISOR project. An IDDP per se however was not tested. Experimental applications of real, long-term IDDPs in a specific water resource / river basin plan will allow testing the strengths and weaknesses of the process, introduce improvements, and identify specific needs for particular natural, political, cultural contexts. Particular emphasis should be given in looking for the most effective combinations of different tools and different representation / participation practices.

Challenge 3 - Improving the use of scientific information in deliberative processes

The management of scientific information was recognised as a key issue in all three ADVISOR deliberative experiments. Whereas deliberation and public participation are necessary in principle for addressing scientific complexity and uncertainty, in practice this ideal meets many problems. Participants in the face of uncertainty tend to defer to perceived scientific expertise or opt for simplistic or emotional solutions. Such problems are exacerbated in contexts (such as those of the southern European cases) where there is a notable lack of even very basic data, not to mention alternative sources of information or expertise.

Post-normal science and the normative guidance of Quality Assurance provide the principles for action. The challenge is to experiment more with these principles in actual decision making processes and to find acceptable and effective ways of combining unconstrained deliberation with scientifically informed debate and decisions.

Challenge 4 – Increasing the decision impact of IDDPs

The theory and "science" of deliberation is one step ahead of its actual practice in real decisions. Deliberative processes have yet to make an impact, other than as experiments, as platforms of civic dialogue, education or at best, social learning. In a few instances, they have been used for consultation processes (e.g. in controversial scientific issues). The challenge is how to make deliberative processes an essential component of democratic decision-making. This challenge is even more pronounced for water resource planning and management, which in many cases is still considered as a highly technical issue, constrained to debates between "experts".

The Opportunity of the Water Framework Directive

The mandate of the WFD for planning, evaluation and participation provides an ideal opportunity for developing, applying and improving IDDPs and linking them to real, acute decision structures. The "ball" is now "in the hands" of the national and river basin authorities. They have to take the next step and attempt to implement deliberative processes in the preparation of the river basin plans, learning from problems and difficulties. The easy solution is to satisfy only the basic requirements of the Directive for consultation (after the decisions), without moving to the active involvement of the public in the actual making of the decisions. Moving to the latter is daring, because it is "unknown terrain" for most public administrations and because it will require some erosion of their power and of traditional channels of influence. Without it however, there are serious doubts whether the goals of the Directive can be achieved and conflict avoided.

Challenge 5 - Changing governance structures

Deliberative processes cannot succeed unless they are embedded in a broader socio-political framework that is itself discursive, defined by deliberation principles and supportive (financially, politically) of deliberative processes. Changing the broader political, social and economic framework towards the ideal of a deliberative (or discursive) democracy requires radical changes which go beyond changes *within* existing political structures.[10] More social science research and analysis is needed in identifying the contours of such a deliberative democracy and its requirements, success and barrier factors and applicability in different socio-cultural or political-historical contexts.

Beyond research however, actual steps and transformations in the political domain are needed. Recent developments in EU policy, such as the White Paper on Governance or in the environmental sphere, the Aarhus Convention and its transposition in Community law, are some first, small steps. More daring steps will be needed in constitutionalising and regulating/institutionalising deliberative democratic processes. In particular clear rules ("entitlements") to deliberate or "represent" should be established and substantial public human and financial resources devoted for deliberation purposes.

The WFD leaves much to be desired in terms of such a shift towards a deliberative water governance. There is no clear definition of the level and type of participation expected, nor clear entitlements on who has the right to "represent" or "participate". Furthermore, there are questions concerning the real leverage of the Directive in comparison to other policy and power structures (e.g. economic, agricultural policies).

Still, it is a first step upon which deliberative applications can build and if successful, gradually contribute to the evolution of broader deliberative water governance.

[10] Dryzek, J.S. (2000), Deliberative Democracy and Beyond. Liberals, Critics, Contestations. New York: Oxford University Press.

The Research Challenge

EU Research has an important role to play in supporting the development of IDDPs in water, and more generally environmental, policy. Key areas of research include:

1. Research and development of modified, new, or hybrid tools.
2. Demonstration of (and learning from) application in actual decision-making cases, especially in the context of the WFD and other EU environmental policies.
3. Inter-comparative research on applicability of tools in different regional/local contexts.
4. Theory of deliberative democracy and discursive governance.

Preferably, research on deliberative processes should be recognised as a separate key theme (for a Network of Excellence or an Integrated project), either in the environmental and/or governance related work-programmes, rather than included in other themes. This will help in creating a "critical mass" of European research in deliberative processes.

Exchange of information between practitioners of deliberative processes around Europe is urgently needed. There is a communication "void" between practitioners working in similar deliberative tools but in different scientific disciplines, or in different countries. The long and considerable experience with local deliberation practices in development studies is not sufficiently taken up by new researchers in environmental policy studies. There is also a relative lack of exchange of expertise between the significant number of deliberation practitioners in the U.K. and the rest of Europe. EU research support actions (including conferences, thematic networks, etc) can provide a useful tool for strengthening such expertise.

Capacity development of future researchers and practitioners is also important. Opportunities for Advanced Study Courses, Marie Curie research grants, researcher exchanges etc., should be fully explored.

There are also several opportunities for International Co-operation in the field. Experience in deliberative processes, especially from North America and Australia where similar research programmes have been developed, could enrich European efforts. Vice versa, European experience could be tested, adapted and learn from different geographical or cultural contexts.

References

Aldred, J. and Jacobs, M. (2000), Citizens and wetlands: evaluating the Ely citizens' jury. *Ecological Economics*, **34** (2), 217 – 232.

Barreira, A. and Kallis, G. (2003), The EU Water Framework Directive and Public Participation in transboundary river basin management". In *The role and use of environmental information in European transboundary river basin management*, (eds. J.G. Timmerman and S. Langaas), pp. 92-107, IWA Publishing, London, UK.

Belt van den, M. (2004), *Mediated Modeling: A System Dynamics Approach to Environmental Consensus Building*, (ed.), Island Press, Washington, DC.

Blamey, R., James, R. and Niemeyer S. (2001), *Citizens Juries for Environmental Management: An Alternative to CBA?*, Final Report, Research Project ANU11 of the Social and Institutional Research Program of Land & Water, Australia.

Bloomfield, D., Collins, K., Fry, C. and Munton, R. (1998), "Deliberative and Inclusionary Processes: their Contribution to Environmental Governance", paper presented at the first ESRC "DIPs in environmental decision-making" Seminar, 17th December 1998.

Bloomfield, D, Collins, K, Fry, C. and R. Munton (2001), Deliberation and inclusion: vehicles for increasing trust in UK public governance?. *Environment and Planning C: Government and Policy*, **19**, 501-513.

Bruisma, F. and R. Vreeker (1994), "Combined Cost Benefit Analysis and Cost Effectiveness Analysis Framework for Flood Protection: Case Study of the Rhine and Meuse River Basins" (available at ADVISOR website: http://ecoman.dcea.fct.unl.pt/projects/advisor).

Commission of the European Communities ComEC (2001) *European Governance: a White Paper*, COM 428, Brussels.

Council of the European Communities (2000), Council Directive of the European Parliament and of the Council establishing a framework for Community action in the field of water policy. Official Journal of the European Communities, L 327.

Coote, A. and J. Lenaghan (1997), *Citizens'Juries: theory into practice*, Institute for Public Policy Research, London.

Corral Quintana, S. (2000). *"Una Metodología Integrada de Exploración de los Procesos de Elaboración de Politicas Públicas"*. Universidad de La Laguna, Tenerife. EC - Joint Research Centre: EUR 19724 ES.

COSLA (2002), *Focussing on Citizens: A Guide to Approaches and Methods*, COSLA, Edinburgh (available also at www.cosla.gov.uk).

Crosby, N. (1995), Citizens Juries: one solution for difficult environmental questions, In Renn, O. et al (eds), *Fairness and Competence in Citizen Participation*. Kluwer Academic Publishers, Dordrecht.

De Marchi et al. (2000), Combining Participative and Institutional Approaches with Multi-criteria Evaluation. An Empirical Study for Water Issues in Troina, Sicily. *Ecological Economics*, **34**, pp. 267-282.

Dryzek, J.S. (1990), *Discursive Democracy: Politics, Policy and Political Science*. Cambridge University Press, Cambridge.

Dryzek, J.S. (2000), *Deliberative Democracy and Beyond. Liberals, Critics, Contestations*. Oxford University Press, New York.

European Commission (2001), *Strategic Document. Common Strategy on the Implementation of the Water Framework Directive*.

European Environment Agency (2001), *Participatory Integrated Assessment Methods*, Technical Report 64, EEA, Copenhagen.

Funtowicz S, and Ravetz J.R. (1990), Uncertainty and Quality in Science for Policy, *Kluwer, Dordrecht*.

Funtowicz, S. (2001). Peer review and quality control. In *International Encyclopedia of the Social & Behavioral Sciences*, (eds, N. J. Smelser & P. B. Baltes), *pp. 11179-11183, Pergamon, Oxford.*

Gibbons, M. (1999), Science's new social contract with society. *Nature* 402, C81-C84.

Global Water Partnership (2000), Integrated Water Resources Management, *TAC Background Papers no 4*, Technical Advisory Committee, Global Water Partnership, Stockholm.

Grigg, N.S. (1996), *Water Resources Management – Principles, Regulations and Cases*, McGraw-Hill, New York.

Guimarães Pereira, A. (2003), "Quality of Information in River Basin Governance, Review of Case Studies: Information Vertex", Report Prepared for the Advisor Project.

Guimarães Pereira, Â., Blasques, J., Corral Quintana, S. & Funtowicz, S. (2003), *TIDDDs - Tools To Inform Debates Dialogues & Deliberations. The GOUVERNe Project at the JRC*. European Commission, Ispra: EUR 21880 EN.

Guimarães Pereira, Â., Corral Quintana, S. (2002), "From Technocratic to Participatory Decision Support Systems: Responding to the New Governance Initiatives", *Journal of Geographic Information and Decision Analysis*, **6** (2), pp. 95-107

Hart, L.B. (1992) *Faultless Facilitation*, Kogan Page, London.

Healey P. (1997), Collaborative *Planning: Shaping Places in Fragmented Societies*, Macmillan, Basingstoke, Hants.

Heron, J. (1993), *Group Facilitation: Theories and Models for Practice*, Kogan Page, London.

Holmes, T. and Scoones, I, (2000), "Participatory Environmental Policy Processes: Experiences from North and South", IDS Working Paper no 113, Institute for Development Studies.

Howell, Johnna L. (1995), *Tools for Facilitating Team Meetings*, Seattle, Integrity Publishing.

IEMA (2002), *Guidelines on Participation in Environmental Decision-Making*, Perspectives Series, Lincoln: Institute of Environmental Management and Assessment)

Institute for Public Policy and Research (1999), *Models of Public Involvement*, IPPR, London.

International Institute of Administrative Sciences (1999), *Citizens and the New Governance - Beyond New Public Management*, IIAS Monographs, Vol. 10, IOS Press, The Netherlands.

Joss, S. (1998), Danish Consensus Conferences as a model of participatory technology assessment, *Science and Public Policy*, 25 (1), 2-22.

JRC (1996): *NAIADE Manual & Tutorial*, Ispra, Joint Research Centre.

Kallis, G. and Butler, D. (2001), The New EU Water Framework Directive: Measures and Implications. *Water Policy*, vol. 3(3), 125-142.

Kaner, S., Lind, L., Toldi, C., Fisk, S. and Berger, D. (1996), *Facilitator's Guide to Participatory Decision-Making*, New Society Publishers, Philadelphia.

Lanz, K. and S.Scheuer (2001), *EEB Handbook on EU Water Policy under the Water Framework Directive*, European Environment Bureau, Brussels.

Munda, G. (1995), *Multicriteria Evaluation in a Fuzzy Environment*, Physica-Verlag, Berlin.Munda, G. (2004): Social multi-criteria evaluation: methodological foundations and operational consequences. *European Journal of Operational Research*, vol. **158** (3), 662-677.

Niemeyer, S. and C. L. Spash (2001), Environmental Valuation Analysis, Public Deliberation and their Pragmatic Syntheses: A Critical Appraisal. *Environment & Planning C: Government & Policy*, **19**(4), 567-586.

O'Neill J., 2001, Representing people, representing nature, representing the world. *Environment and Planning C: Government and Policy*, **19**(4), 483-500.

Okubo D. (1997), *The Community Visioning and Strategic Planning Handbook*, CO, The National Civic League. Denver.

Pimbert, M.P. and Wakeford, T. (2001), Deliberative democracy and citizen empowerment, Special Issue on PLA (Participatory Learning and Action) Notes, 40, February 2001,International Institute for Environment and Development, London.

Shipley R, (1997), Visioning in planning: is the practice based on sound theory? *Environment and Planning A*, **34** 7-22.

Spash, C. and K.Urama (2003), "Monetary Valuation Methods Reviewed for Relevance to River Basin Management", as part of the Advisor project (available at Advisor website: http://ecoman.dcea.fct.unl.pt/projects/advisor).

Stave, K. (2002), Using system dynamics to improve public participation in environmental decisions. *System Dynamics Review*, **18** (2), 139-167.

Street P. (1997), Scenario workshops: a participatory approach to sustainable urban living?.*Futures* **29** (2) 139-158.

Vennix, J. (1996), *Group Model-Building: Facilitating Team Learning Using System Dynamics*, JohnWiley & Sons, Chichester.

Videira, N., Antunes, P., Santos, R., Gamito, S. (2003) Participatory Modelling in Environmental Decision-Making: The Ria Formosa Natural Park Case Study. *Journal of Environmental Policy and Management*, **5**(3), 421-447.

Weisbord M., Janoff S. (2000), *Future search. An action guide to finding common ground in organizations and communities*, Berrett-Koehler Publishers, San Francisco.

Wong, A.K. (1999), Effective Public Participation in the Rate Setting Process: LADWP Blue Ribbon Committee on Rates, (eds. Gleick, P.H and A.K. Wong), *Sustainable Use of Water California Success Stories*, Pacific Institute, Oakland.

IWA Publishing's authorised EU representative for General Product Safety Regulations is Diane D'Arras, 15 rue Duret, 75116 Paris, France, e-mail: safety@iwap.co.uk.

Printed and bound by CPI Group (UK) Ltd, Croydon, CR0 4YY
12/05/2026
02108874-0016